A Brief History of Scottish Beekeeping and Beekeepers

by

Taylor Hood BSc PhD

A Brief History of Scottish Beekeeping and Beekeepers
© Taylor Hood BSc PhD 2024

Published in the United Kingdom by
Northern Bee Books,
Scout Bottom Farm,
Mytholmroyd,
West Yorkshire HX7 5JS
Tel: 01422 882751
Email: jerry@northernbeebooks.co.uk

www.northernbeebooks.co.uk

ISBN 978-1-914934-92-6

Design and artwork: DM Design and Print

A Brief History of Scottish Beekeeping and Beekeepers

by

Taylor Hood BSc PhD

Contents

Acknowledgements	p 1
Foreword	p 3
The Origins of the Hive	p 5
John Gedde	p 8
James Bonner	p 15
Robert Kerr and the Stewarton Hive	p 19
The Walker Family	p 29
John McPhedran	p 31
William Thomson	p 33
William Raitt	p 39
Miss Clementina Stirling Graham	p 42
Alexander Pettigrew	p 44
William McNally	p 50
George Avery and The Scottish Beekeepers' Association	p 52
Richard Whyte	p 57
John Wilson Moir	p 63
D M MacDonald	p 67
Dr John Anderson	p 70
Dr Edward Jeffree	p 75
Alex S C Deans	p 77
Dr Delia Allen	p 83
Margaret (Peggy) Logan	p 87
Alex R Cumming	p 88
Bernhard Mobus	p 90
James Savage	p 93
Captain L M Thake	p 100
Mr & Mrs Shepherd	p 105
Willie Smith	p 111
William Hamilton	p 118
Braithwaite Brothers - James and George	p 122
Robert Couston	p 131
The Glasgow Beekeepers -	
Ian Craig, Eric McArthur and Charles Irwin	p 135
Andrew Abrahams	p 139

Acknowledgements

I would like to thank Michelle Berry, Enid Brown, Phil McAnespie, Alan Riach, and Bron Wright of the Scottish Beekeepers' Association for their help in sending me information, lending me materials, books and magazine articles, including some from the Moir Library, which I have used in the writing of this Book.

I would also like to thank Malcolm Watson of Aberdeen Beekeepers' Association and William Hunter of the East of Scotland Beekeepers' Association for information and photographs on Delia Seager and James and George Braithwaite respectively.

A Brief History of Scottish Beekeeping and Beekeepers

Foreword

Our actions are usually influenced by what we see/hear and what we want. When this is to do with bees, what we want and what the bees want to do may be very different and this difference can have consequences that may be fatal to the bees.

It was Bernhard Mobus who said "dead bees don't produce honey" and Captain Thake who said that "when keeping honey-bees you should follow the natural cycle of bees".

The natural cycle for bees is primarily one of colony survival through the winter, followed by the colony prospering in the Spring and then to colony reproduction, usually in late Spring early summer. Hooper in his book *Guide to Bees and Honey* 5th Edition (p 51 and p 101) put the beekeeper's job through the season as 'to assist the broad pattern of the annual bee colony cycle...' by assisting rapid build-up in the spring, holding them together during peak brood rearing (i.e. stopping them from swarming), ensuring they have sufficient stores to last through the winter and to combat pests and diseases.

Sometimes the primary focus of the beekeeper can be different to that of the bees where survival is their main focus. The beekeepers primary focus can be honey production, queen rearing, producing nucleus of bees to sell etc. and this focus can sometimes override the focus on bee survival. This is sometimes through ignorance, following bad practice, or perhaps even greed or penny pinching. This difference in focus can therefore be to the detriment of the bees. I know of some beekeepers who take off every ounce of honey they can in September, making up too many splits when the bees produce queen cells and even not treating for varroa properly. All these factors/ actions can be fatal to the bee colonies involved.

Willie Robson of Chainbridge told me "bees want to be immortal" their instincts are of survival and yet so many beekeepers rather than working with the bees fight against their natural cycle and behaviour and therefore are not working with the bees to survive and prosper.

We can learn much from the past, particularly through the development and successes of past expert beekeepers.

Willie Smith did not feed his bees sugar syrup unless he had no other alternative – he only took honey that was surplus to the bee colonies need, bee nutrition was important to him. What impact does, removing all the colonies honey and replacing it with some sugar syrup. I am sure it affects their temperament (more aggressive), bee morale (again something Willie Robson talked about) their motivation to live as well as their ability to fight off pests and diseases and perhaps poor larval and bee health due to poor nutrition and the consequences that this brings.

I have heard of beekeepers on finding queen cells in a colony producing three or four nucleus colonies rather than just one and then later on in the season when preparing for winter trying to unite them all together as there were not sufficient bees in the colonies to survive even a moderate winter.

All beekeepers need a good form of swarm prevention and control and where better than seeking the advice of a successful expert beekeeper.

To control varroa we use miticides / and other chemicals. Over a short period of time they are effective in killing the varroa mite and helping the bees survive the impact of the varroa infestation but long term what effect do these miticides have on bee / queen health. The varroa can become resistant to certain miticides, the chemicals are accumulating in the wax comb with the fertility of drones being affected and therefore long-term effectiveness of queens and colony survival.

We learn from our mistakes but what happens if we keep on making the same mistake and don't even realise we are making a mistake(s), continually blaming the bees and the environment we live in, rather than looking at ourselves and our beekeeping practices.

As a group, beekeepers need to review what they are doing and reassess their practices so that they reflect the natural cycle of the honeybee and only use treatments and practices they would be happy to treat themselves with.

I have written this book on Beekeeping and expert beekeepers, to show how beekeeping has developed and progressed over time with the practice and thinking of some great beekeepers. I hope it makes the information about them more accessible and makes those beekeepers who read the book, challenge their current beekeeping practices and it makes them better beekeepers. I know it has for me.

The Origins of Bee Hives

Honey bees are naturally wild, nesting in colonies in small cavities e.g. in rocks or trees which have small openings which can easily be defended / guarded, which are dry, well ventilated and have adequate room for the colony to breed and store food to survive times of dearth.

Humans have been keeping bees for millennia. Initially man /woman were honey hunters who sought out honey bee colonies and removed honey comb for food. This still happens in some countries e.g. Nepal. Honey hunting progressed from natural hives in trees, to door sections on tree trunks being cut away to allow access to the honey combs which could then be removed, harvested and the honey and wax extracted. The door could then be replaced allowing the bees to rebuild the wax comb and to continue to collect stores. Later as hand saws became available, trees could then be cut above and below the honey bee nest and the part containing the honey bees and their nest could be moved to a more convenient spot, probably to a place more accessible to human beings.

The first hives that were constructed by humans were based on natural honey bee nest sites with materials that were readily available e.g. clay pipes in Egypt, wood, wicker and straw in Northern Europe. It was from Germany that wicker /straw skeps were made and used to keep bees. This practice then spread through the rest of Northern Europe including Britain. This all happened before the Romans arrived in Britain in 55BC. These straw skeps / small conical wicker hives were "cloomed" with dung and lime and covered with hackles to make them resistant to rain and free from draughts. These hives are still called ruskies derived from the words Rusca /Ruschae meaning wicker hive.

During mediaeval times it was the church that kept beekeeping alive. Candles from beeswax were required to light catacombs as well as becoming part of liturgical services and celebrating Mass. Candles were also being lit to celebrate a person's momentous Christian life milestones – at baptism, communion, wedding day and death.

On Easter day a candle was lit to symbolise the lighting of the shadows and it was kept burning until Ascension Day. Beeswax was described as a " pure product provided by virgin queens, formed from the most exquisite essence of flowers".

Bees were seen by the Church as a symbol for Christian virtues and of Christian morality in all aspects. The Church endowed the bees with patron Saints.

Even to this day Monks are keen and skilled beekeepers.

Beekeeping and hive making did not change from prehistoric times until the 16th Century with the advent of printing when information on many subjects became available and the development of many areas/ things including beekeeping began to happen.

Initially it was translations of Virgil and Aristotle that were published on beekeeping. The first British book entirely on bee keeping was published in 1593 and was written by Edward Southerne. Charles Butler's book *The Feminine Monarchie* was published in 1609.

In 1675 John Gedde published his treatise on beekeeping. This book was seen as a considerable advance in beekeeping, building on the work of others including Purchas, Levett and Butler.

In 1673 Sir William Thomson sent the design of a hive to Gresham College of a Hive "used in Scotland with good success". It was a tiered hive - a rectangular wooden box with a straw skep above it.

Fig 1.1 La Ruche Ecossaise The Scotch Hive

Its origin or when it was designed was not recorded however French beekeepers described the hive, its use and adaption e.g.Bourdonnaie, Docouedic and de Gelieu. Docouedic added another box to the hive and it was then known as the Pyramidal Hive. This hive was seen as a precursor for the hive designed by Warre.

Many of these early books can be found in the Moir Library in Edinburgh.

One of the earlier English Books with a section on Beekeeping in the Moir collection is by Samuel Hartlib : his *legacie; or, An enlargement of the discourse of husbandry* 2nd ed., London 1652.

Samuel Hartlib was Polish and came to Britain in 1628. He died in 1670. He wrote *The Reformed Commonwealth of bees* in 1655. (His book was written during Cromwell's time therefore the name)

In this book he describes two hives one by the Rev. Dr. Brown and the other by the Rev. Will Mewe. It is interesting that through the thinking at that time nadiring was preferred to supering and round hives were preferred to square hives as it was thought to be more in keeping with what the bees did / preferred in nature - bees building from top down and bees building their nests in trees.

The hive of Brown was made of circular bodies with iron handles made from the wood of casks with the capacity of a bushel (Space equivalent to 36.4 litres) although some were made square with 4 boards, but round was preferred and seen as the most suitable for bees. Three of those brood boxes (bodies made up the hive). Since the bees were not able to work large void spaces easily without support, a structure of gardeners' frames composed of posts and hoops with cross bars at the top and the middle were proposed by Brown. This structure was placed in such a way that the bars could not be shaken or moved thus supporting the honey/ wax comb. This was the first rudimentary bar frame hive. It was not until 1780 that John Keys designed his wooden hive with 6 parallel bars which were rigidly connected.

The second hive Hartlib described was that of Rev. Will Mewe of Easlington, Gloucestershire and was called the transparent Bee Hive as it had a single glass window on each body box to allow for the comb to be seen. The hive like that of Brown's was a three-tiered hive.

Each brood body had an entrance made up of 3 small openings along with a sliding door that could be opened or closed. Each box had a hole at the top which could be opened or closed allowing bees to move or not move from one body box to another. The Brood body boxes were lined with rush matting and the hive had an outer casing. Christopher Wren, Fellow of All Soules College, Oxford wrote on February 26th, 1654 (When 23 years of age and not famous) of his experience with the hive.

The Gedde / Gresham Hive is a combination of the Brown and the Mewe hives along with modifications made by Gedde.

John Gedde
(1647 - 1697) and his Octagonal Hive.

In 1668, John Gedde of Falkland, Fife designed his Octagonal hive. It has been written that the design was based on his observations of wild honey bee colonies in oak trees at Falkland Park. The description of his hive was published on July 21st, 1673 in number 96 of the Royal Society's Philosophical Transactions and the hive was exhibited at Gresham College, London. The Royal Society was founded at Gresham College in 1660 and was based there until 1710.

Gedde's hive was not the first octagonal hive, that honour goes to the Rev. W. Mewe of Gloucestershire. In 1675 Gedde published his *Treatise on Beekeeping*: -**A new discovery of an excellent method of bee-houses, and colonies, which frees the owners from the great charge and trouble that attends the swarming of bees, and delivers the bees from the evil reward of ruine, for the benefit they brought their masters**. Gedde obtained a patent for his hive for 14 years from King Charles II. This was the first bee hive ever to be patented,

King Charles II set up apiaries at Spring Gardens in Whitehall and in Windsor using Gedde hives and took pleasure in watching the bees at work and the honey being removed without disruption to the bees.

The King arranged for Gedde to get 20 acres of moorland at Falkland Park and to receive £200 for his work and to stir up interest in beekeeping.

Gedde and his partners travelled through the English counties promoting the use and the purchase of his hive.

In 1685 on the death of Charles II, John Gedde was able to renew his patent for his hive with the new King - James VII Scotland, James II Britain. However, John Gedde did not comply with the conditions and oaths that were required by the patent and was therefore pursued by the crown. John Gedde had to leave the country until the accession to the Throne by William of Orange.

In 1697 Gedde published an enlarged version of his book on bees. The last edition (the fourth edition) was published in 1721.

Gedde was a good beekeeper and believed it was foolish to kill the bees to get honey. He believed that swarming was caused by a lack of space and that swarming was not something you wanted to promote unless you wanted to make increase. Gedde believed his hive was the most productive and therefore profitable hive available at that time.

The Hive

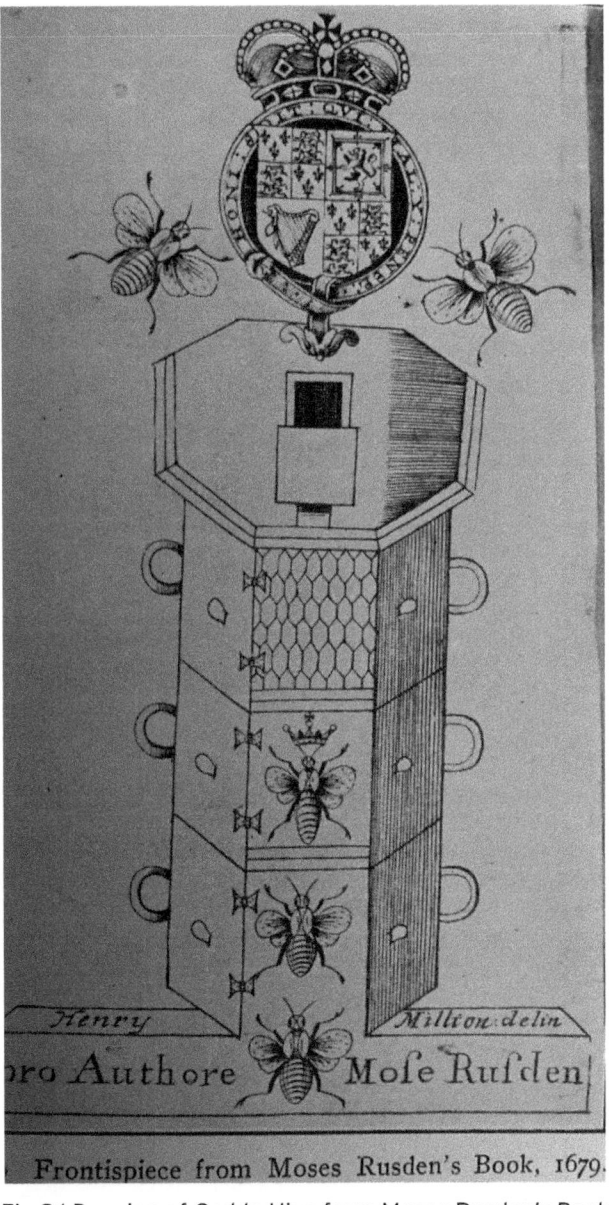

Fig 2.1 Drawing of Gedde Hive from Moses Rusden's Book

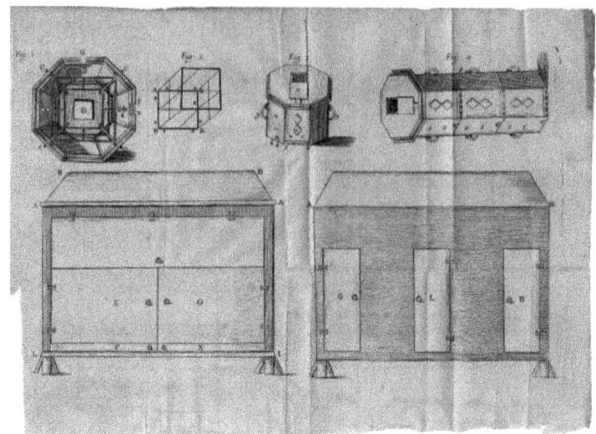

John Gedde. *A new discovery of an excellent method of bee houses & colonies ...* 1675. Folger

Fig 2.2 External and Internal Drawing of Gedde Hive

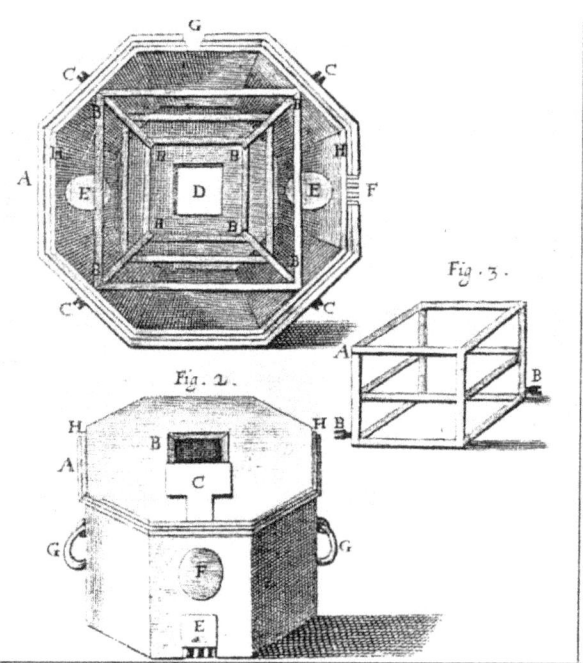

Fig 2.3 Drawings of Gedde Chamber and internal frame

The hive consisted of chambers / compartments (usually sold with 3 chambers) which could be added or removed as necessary to meet the needs of the colony. The chambers were octagonal, 12 inches deep and had a breadth of 16 inches (the breadth was a third more than the depth), the octagonal face was 9 inches wide.

Within each chamber was a wooden square frame which was attached to the side of the chamber. It was on this frame that the bees built their wax comb. The top of the chamber which was covered had a 4 inches square hole with slide so that the chamber could be closed or opened as required. There were two diamond shaped panes of glass, back and front of the chamber to allow for observation to check the size of the colony and whether there was enough space for the bees. The beekeeper could then decide whether or not he wanted to add or take away chambers as necessary.

Empty chambers were usually nadired under a full chamber.

Handles to allow for easy lifting of the chamber were positioned in the middle of the left and right sides of the chamber.

The entrance consisted of 6 openings ½ an inch wide and 1 inch in height, with each opening having a slide shutter which could be opened or closed.

The chamber boxes had a rebbit ½ inch in depth on the outside on the top and another rebbit on the inside of the bottom 1 inch in depth so chambers could fit together snuggly.

The Frame was made up of 4 posts and 19 small sticks.

12 sticks were fastened to the posts.

6 sticks which crossed the sticks fastened to the posts.

1 stick attached to middle cross sticks in the centre of the frame which allowed bees to ascend and descend between the chambers.

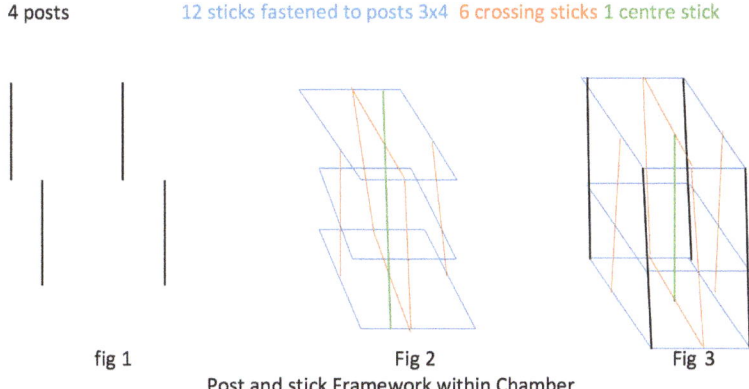

fig 1   Fig 2   Fig 3
Post and stick Framework within Chamber

Fig 2.4 Gedde Hive Frame work.

(Fig 2 The central stick in the diagram is longer than it would have been - it would have gone from the top cross stick to the bottom cross stick as in Fig 3.)

Populating a Gedde Hive.

The hive could be populated in two ways.

1. A swarm could be introduced directly into a chamber
2. A skep of bees could be placed on the top of the hive/chamber using an adaptor board, allowing the bees to go into the chamber below. When the chamber was seen to be full of comb another empty chamber was nadired below.

Harvesting honey

Using the small access hole on the bottom of the upper chamber the slide on the lower chamber was drawn over the hole in the bottom chamber. The top chamber was then left for 30 minutes before lifting the top chamber up slightly and held open with a small wooden block. The bees which had been unable to escape realising they could now do so, would quickly leave the chamber and make their way back to the colony in the lower brood chamber. When no more bees were seen to leave the chamber, it was taken away to a suitable place, turned upside down, covered with a cloth and left overnight. The next morning the cloth cover was removed allowing any bees still in the chamber to return to the hive. The honey could then be harvested. This was done by unscrewing the frame from the chamber and then taken out. As Gedde put it " unscrew the pins and let out the frame with the whole fabric in which there were no bees, harvest what you think the bees can spare, screw the frame back, open the slide on the box below and put the top box back. This provides a reserve for the bees in winter. If the conditions are right the whole top box can be removed and another fresh box added below."

It is interesting to note that the frame after the posts were unscrewed from the chamber could be removed i.e. a moveable frame. I am not sure how easy this would have been and it is unfortunate that nothing has been written about how the bees built/ attached the wax comb to the frame. The hive was not managed through top bars as I originally had thought or by the bees attaching comb to the roof of the chamber. I now wonder how the combs were configured/spaced by the bees on the frame and why no or very little propolis or burr comb was attached to the chamber body which would have made it very difficult to remove the frame.

A number of octagonal hives were kept in a bee house/ cupboard to better insulate the hives and to protect the bees . This limited the number of hives that could be kept and made it difficult to manipulate the hives.

Fig 2.5 Drawing of Gedde Beehouse - outside view

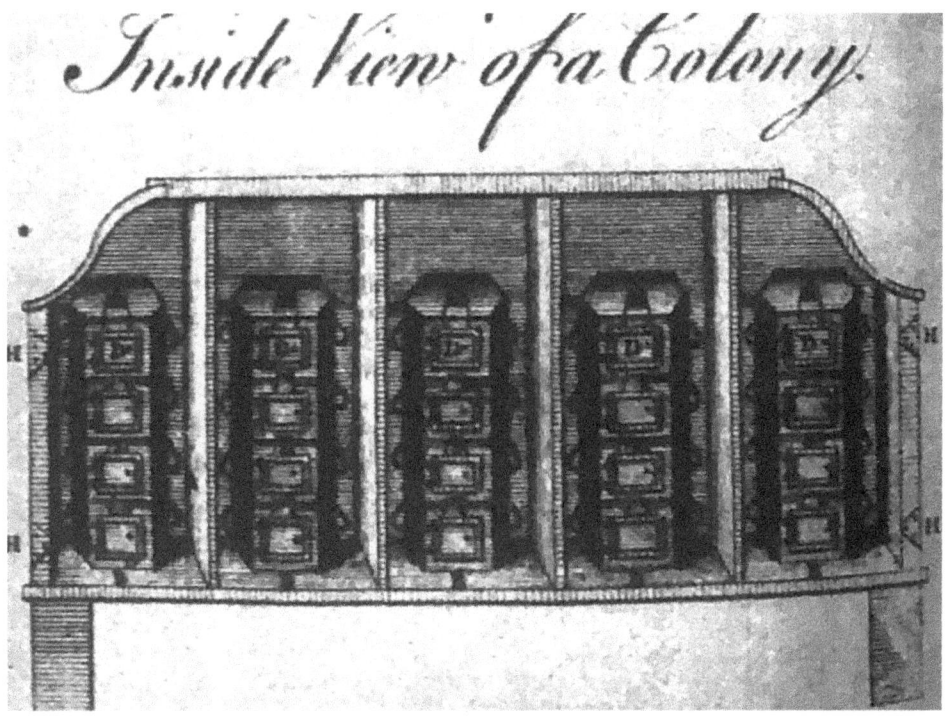

Fig 2.6 Drawing of Gedde Beehouse inside view

It is evident that Gedde cared for his bees' survival more than the profit he could make from taking all their honey.

Gedde was a successful beekeeper and showed that bees could be kept productively and profitably without the need to kill them. There is much more to his hive than just 3 octagonal boxes. However, the manipulations required to work the hive may have been quite difficult and thus did not revolutionise the beekeeping of that time as Gedde had anticipated and therefore his dream of fame and fortune was not realised.

A notable beekeeper who followed Gedde was James Bonner who was possibly the First Commercial Beekeeper in Scotland.

James Bonner

James Bonner came from Auchencrow near Berwick upon Tweed. In 1789 he published the first of his books on beekeeping – *The Bee-masters Companion and Assistant*. In 1795 he wrote his second book on beekeeping – *A New Plan for Speedily Increasing the Number of Beehives in Scotland* – a large part being a repetition of his first book.

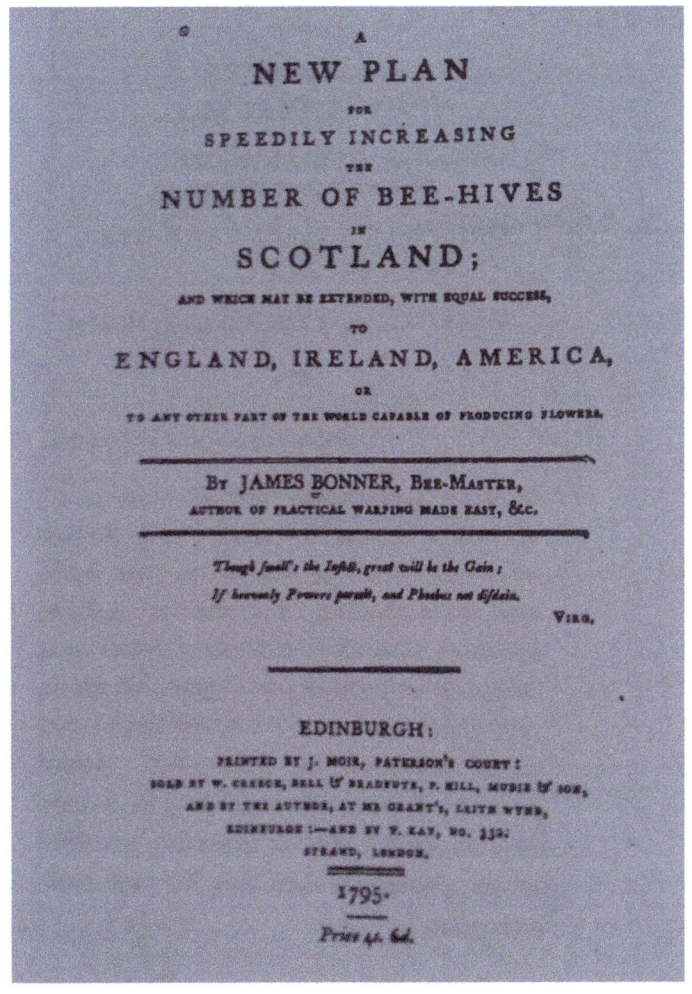

Fig. 2.7 Bonner Book title page

In the preface of his books he tells us that he was interested in bees from an early age being employed to watch his father's bees during swarming time and that he felt "in a kind of paradise when running among the bushes and seeing the bees swarm". He purchased 3 hives at a young age (16 Years) and gradually increased his stocks over the years. He avidly read every book on beekeeping he could get his hands on, as well as actively learning as much about them from his own observations.

Bonner tells his readers that bees " are well worth our care and attention ....and when properly managed bring their owners yearly a considerable profit with no rent or tax and with little attendance." He sites that the principle reason that honey bees are not widely kept is the lack of knowledge people have about them and how to look after them. Much of what he wrote is as relevant today as when he wrote his books although as he said in his books – "beekeeping knowledge is in its infancy and that some mistakes have been made."

Bonner dedicated his second book to the Most Noble, The President, and the other Right Honourable and Honourable Members of the Highland Society of Scotland, due the patronage they had given to the writing of the book as well as in his commercial concerns as a dealer in honey. James wrote about meeting Sir John Sinclair one autumn morning in 1794 when he was delivering honey combs to a gentleman in the New Town of Edinburgh. Sir John asked him to supply him with some honey comb the next day. When delivering the honey, a conversation between the two took place where Sir John asked Bonner to write a plan for rearing bees and producing honey in Scotland in a more extensive manner. Sir John then presented the plan to the Highland Society which gave Bonner the patronage/ subscriptions to allow him to present his plan to the wider public through his book.

Sir John Sinclair of Ulbster, Caithness was considered to be one of the greatest living Scots in the late 18th Century, he studied at the Universities of Glasgow, Edinburgh and Oxford, was a member of the Bar in both Scotland and England and was the founder of the Scottish Board of Agriculture and its President for 13 years. From his meeting with Bonner - much of its content was recorded in a letter written from Bonner and to Sinclair in September 1794. The letter was published and can be read in its entirety in the August 1932 *Scottish Beekeeper*. In this letter he talks about having been a beekeeper for more than 40 years and from his beekeeping had been able to support a wife and ten children from the bees' produce. At that time, he kept 80 hives. In the letter he highlighted the purchasing of honey made from other countries. He believed that by increasing the number of hives in Scotland that much of the honey bought could be supplied by those additional Scottish hives and that the honey would also be of superior quality.

His plan was to increase the number of hives in Scotland by at least 10 to 20 times. He calculated that with 873 country parishes in Scotland with a number of 30 hives per parish that there would be around 26,190 hives in Scotland at that time. He suggested that by increasing to 300 hives per parish there would be 261,900 hives which then could be doubled the following year with an annual increase in Scottish wealth per year of £261,900 (with the value of £1 per hive) as well as the money generated by the vast amounts of nectar collected and honey produced from this number of hives. He believed that there was sufficient forage in Scotland to meet this increase in hives/colonies. He suggested the employment of skilled beekeepers to act as examples and also to superintend (inspect and mentor) until the new beekeepers had sufficient knowledge to keep bees. These people would be given recompense for their work and he also suggested that people who kept bees would be paid a premium for keeping a larger number of hives of bees thus supporting the benefits that would be gained from keeping an increased number of honey bee colonies in Scotland. He suggested that if he took on the role in this plan, looking after Berwickshire, the County of Peebles, the 3 Lothians, Clackmannanshire and Fife he should receive £50 per year and 1s 6d per hive inspected. Also, if he was involved in the training of skilled Beekeepers to superintend he would want £10 per year per beekeeper for training (which would take 2 years) and £10 each for their yearly board. An average yearly salary at that time was around £20.

He saw his book as a treatise on beekeeping, that gave directions on the proper management of beekeeping that would benefit and instruct new and less experienced beekeepers.

So, what about James Bonner the beekeeper.

Robert Huish from Nottingham an educated man, beekeeper and author of: *Bees - Their Natural History and General Management* - knew James Bonner and wrote about visiting Bass Island with Bonner and observing bees with him. They must have visited hives somewhere else as there were no hives on Bass Island. Huish wrote about how Bonner seemed to be sting proof, an observation he made when he saw Bonner's bald head with multiple stings with no swelling and apparently no pain. Huish wrote that if " any other person experienced a 20th part of these stings, death would probably would have been the result." The only beekeeper I have seen like that is Charles Irwin. Sandy Cran and I were assisting Charlie collect a swarm in a tenement flat in Glasgow which was in the rafters using his collection boxes and a Vax vacuum cleaner. Charlie got stung on the eye lid. Not saying at the time, it was only when he was leaving the building he asked Sandy to remove the sting from his eyelid, which Sandy did. At the time I thought that when I saw Charlie, the next day that his eye would be so swollen and that he would not be able to see out of it. The next day when I saw Charlie he was just as he usually was, no swelling just normal.

Bonner's advice in working with bees was "In all things you have to do with bees, do it in a soft, calm, gentle submissive manner." A reminder to us as the song goes - "it's not what you do but the way that you do it" . Beekeeping is an art and is skill based and when it comes to working bees, things have not changed much over the years. No wonder James Bonner was called the Beemaster of Auchencrow.

Huish described the Ruche Ecossaise in his book and its management. Huish described La Ruche Ecossaise as a hive recommended by Bonner.

Another notable beekeeper of that time was Robert Kerr - Bee Robin who designed the Stewarton Hive.

Robert Kerr and the Stewarton Hive.

The Stewarton Hive is believed to be the first standardised hive ever made. Robert Kerr (Bee Robin) of Stewarton, a cabinet maker and keen beekeeper produced the hive for himself. He tried different shapes and sizes and in 1819 produced the hive we now know as the Stewarton Hive (Also known as the Ayrshire Hive).

The hive was an improvement from skeps and box hives – it was based on the octagonal hives of the Rev. William Mewe, Easlington, (Eastington), Gloucestershire, 1652 and that of John Gedde of Falkland, Fife 1675.

Fig 3.1 Stewarton Hive outer cover - display Kelvingrove Museum

Fig. 3.2 Old Stewarton Hive outer cover - Stewarton Museum

The standard Stewarton Hive was made up of three brood boxes (called at that time a breeding or body box) and 4 supers (called honey boxes).

The Brood boxes were 14 inches wide and 6 to 8 inches deep. They were interchangeable and the boxes dove tailed into each other. Each box contained nine - $1^{1}/_{8}$th inch bars which were screwed down with $^{1}/_{2}$ inch screws, and moveable wooden slides $^{3}/_{8}^{th}$ inch wide, could be inserted into groves made in the top bars.

The top bars were placed $^{3}/_{8}$ inch apart, with the wooden slides being put in the spaces, this showed that Kerr had understood the importance of the bee space.

Fig 3.3 Stewarton Hive - 2 inner brood boxes

At the back and front of each chamber was a glass window with wooden shutter to allow the beekeeper to access / monitor the development of the bee colony within. Each box had a 3½ inch entrance.

Through observation, adding brood boxes and supers and properly manipulating the hive, the beekeeper could ensure that the queen had sufficient room to lay in the centre of the hive. The queen rarely visited the outer combs of the brood chambers and therefore almost never moved up into the supers/ honey boxes to lay.

To ensure that the brood nest was compact a double brood body/box system was used as wells as ensuring the 2 combs on both sides of the brood nest were full of honey and pollen. The first super /honey box was not added until this was the case along with the hive being full of bees. When the super /honey box was put on the brood body, the two outside slides of the brood body touching the walls were withdrawn. This allowed the bees to work on the outside combs of the super /honey box. When these combs were half full the next pair of slides were pulled out. The third pair of slides were only pulled out for short periods during high nectar flow and as the flow reduced the slides were inserted again. The honey / combs being produced and stored from the sides into the centre of the super/honey boxes. When the central combs of the super/honey boxes were touching the glass observation window another super honey box was placed on the hive above the super and the slides of the lower super were removed, a pair at a time from the outer sides of the hive until the new super was filled with honey.

When the second super was put on the hive an extra brood box was nadired under the original brood chambers giving the queen and bees extra room to lay and hangout respectively thus reducing the chance of swarming.

Fig 3.4. Stewarton honey box

The super or as it was known at the time, honey box was either $3\frac{1}{2}$ or 4 inches deep. Each of these boxes held 7 bars which were $1\frac{1}{2}$ inches in breadth along with the $\frac{3}{8}$ inch wide slides. The $3\frac{1}{2}$ inch honey box held approximately 14lbs and the 4 inch around 20 lbs of honey. Bees were only allowed access to the supers from the side combs and it was suggested that this stopped the vapours from the brood nest from getting up to the super and staining the white honey comb as well as stopping the bees from walking directly up the combs and causing travel staining.

Stewarton Hives were extensively used to take bees to the heather as it was very easy to transport them.

Some Stewarton hives like the WBC hive of today had an outer case.

In 1874 at the Honey Show at Crystal Palace six Scottish Beekeepers won first prize for their super/honey box of comb honey. The winners that day included James Anderson from Dalry, William Sword from Falkirk and Alexander Ferguson from Stewarton all using Stewarton hives. The supers had beautiful white honey comb, free from brood, cocoons and other products from brood rearing.

Fig 3.5 Super / honey box with white honey comb

Fig 3.6 honey box with white honey comb

Fig 3.7 Honey comb from Stewarton hive

This resulted in a huge interest in the Stewarton hive at that time throughout the U.K.

The Stewarton hive if managed properly produced big crops of honey – around 200lbs in some cases and the bees seldom swarmed. "Bumblebee Johnnie"

Fig 3.8 Photo of Johnnie Walker (Bumblebee Johnnie)

The grandson of William Walker in a talk in 1937 to Glasgow and District Beekeepers' Association said he never got as high a yield of honey from a modern hive as he did from a Stewarton. In fact, John Walker his father who due to illness had to give up his job as a cobbler was able to produce sufficient honey with the help of Johnnie from around 90 colonies in Stewarton Hives to sustain his family for over 20 years. Unfortunately, many of these hives were destroyed, burned because they were thought to be infected by the Isle of Wight disease.

The Stewarton demise came with the demand for extracted honey in jars and section honey. Sections first became available in this country in 1878 when they were imported from America by Mr William Raitt and Mr Robert Steele. The original sections produced 2 lb sections and caused a sensation at that time.

There were at least 3 types of Stewarton Hive, the original as made by Robert Kerr.

The Renfrewshire-Stewarton was a modification of the Stewarton hive made by McPhedran - (known as the Renfrewshire Beekeeper) and a cabinet maker from Stewarton who made Stewarton Hives at that time - James Allan. The modifications of the Renfrewshire-Stewarton hive were the body box increased in depth to 9 inches, the 9 top bars in the brood body were reduced to 8 with the four central bars being replaced with four moveable frames with the four outer bars still fixed with screws, the entrance was increased to 5 inches. Wooden slides similar to the original hive were used.

Fig 3.9.1 Photos of a modified Renfrewshire-Stewarton Hive

Fig 3.9.2 Modified Renfrewshire-Stewarton Hive with 4 national frames

Fig 3.9.3 Modified Renfrewshire-Stewarton Hive with plastic cover rather than slides to restrict bee access to honey boxes to gaps between outside bars.

Fig 3.9.4 Modified Renfrewshire-Stewarton Hive Showing comb on outer bars attached to side of hive restricting bee and particular queen bee movement and access to upper boxes.

The Third, is the Carr-Stewarton Hive was a square hive based on the design of the original Stewarton Hive and designed by Mr C W Smith.

Fig 3.10 Drawing of Carr-Stewarton Hive

The Brood Chambers were 15" square and 6 " Deep and had 9 moveable frames where the top bar was wedged shaped. Wooden slides /guides were included in the design along with windows back and front. The supers were 4 " deep and had 7 bars.

A crown board with slats and slides was also included with the hive.

A design probably copied from that of William Thomson who wrote about this type of crown board in the 1870s in the British Bee Journal.

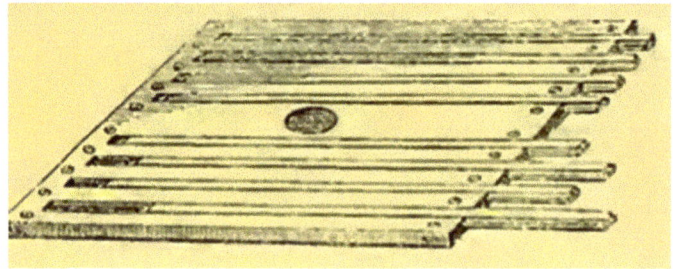

Fig 3.11

Crownboard designed by William Thomson developed from original Stewarton Hive.

The Walker Family

William Walker had helped Robert Kerr (Bee Robin) in the construction of the Stewarton /Ayrshire Hive. This was the first hive to separate the honey from the brood, it could be expanded or contracted as required, was probably the first wooden hive to be used in transitory beekeeping when hives were taken from Ayrshire to Arran for the heather.

The Walker family was a great dynasty of beekeepers - William the grandfather

produced moveable frames for the hive - the spacing of these frames respected the bee space allowing the bees to move freely, not clogging up the hive with burr comb.

Figs 3.12 Moveable frames for Stewarton Hive

When moveable frames were put in an original hive the natural barrier of honey frames and the wax which held the honey comb to the outside of the brood box (See fig 3.9.4) was removed allowing the queen access into the honey supers. William or perhaps John his son produced slides with appropriate gaps to exclude the queen from the honey boxes.

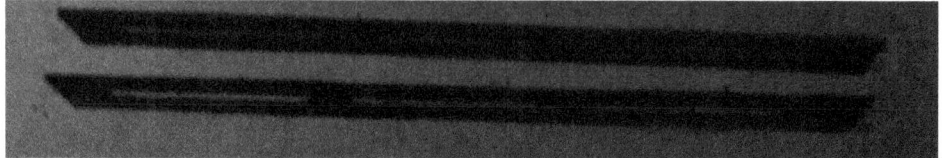

Fig 3.13 Slides used in Stewarton hive along with moveable frames

Fig 3.14 Photo of Stewarton Brood Box showing original Queen Excluder slides
(by James Struthers)

These modifications/changes to the hive were made before 1850 and showed an understanding of the bee space.

John who had been a cobbler had to give up work due to ill health. He with the help of his son Johnnie kept over a 100 hives and were able to sustain the family with the honey they produced.

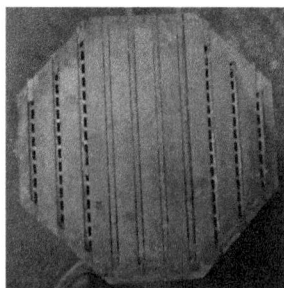

Fig 3.15 Original Stewarton Hive Brood box with normal slides and Queen Excluder slides.
Stewarton Museum

John McPhedran - A Renfrewshire Beekeeper and the Renfrewshire-Stewarton Hive.

John McPhedran was born 1827 and died 1891 aged 63.

He retired to Craigbet in 1854 when 27 and led the life of a country gentleman when he inherited Craigbet. He kept Ayrshire dairy cows, Leicester Sheep and bees and was a great advocate of the Stewarton Hive and its management system.

He was one of the leading beekeepers of his day writing many articles on the subject of bees and beekeeping.

Following the example of one of his neighbours he set up a hive from a colony that had set up home in the roof of his house. On a visit to Glasgow by chance he saw a wonderful display of honey supers (honey boxes) in a grocer's window. On investigation he found that they were supers of clover honey from Ayrshire. He arranged to go to Ayrshire to find out how the beekeepers had produced such fine honey. He then bought a Stewarton hive and worked it as instructed. He was successful in producing a good yield of honey.

He then saw ways in which the hive could be improved and therefore made modifications - calling this hive the Renfrewshire-Stewarton.

The Brood Body was 9 inches deep, had 8 top bars/frames, the central 4 were frames which were moveable and therefore interchangeable, the 4 outer bars were fixed with screws.

The hive had slides and it worked along the same principles of the original Stewarton Hive.

He designed other bee appliances - his queen introduction cage winning 1st prize at the 1874 show at Crystal Palace.

From 1860 he wrote in the Cottage Gardener until the death of Thomas Woodbury in 1870. He imported one of Woodbury's Ligurian queens (Woodbury introduced Italian Bees into Britain in 1859) probably the first in Scotland. Unfortunately, he also imported foul brood from Devonshire, from the bees he got from Woodbury.

McPhedran was introduced through his friend Duncan Keir to Alfred Neighbours and through this association gave Neighbours advice on how to pack and send Humble (Bumble) Bees to New Zealand successfully to pollinate red clover.

McPhedran also supported John Lowe of Dunkeld about the cause and cure of foul brood.

John Lowe wrote that foul brood was not an artificial disease caused by experimentation as was thought at the time and John McPhedran supported him on this matter.

John Lowe was a banker - born 18th May 1813 and died December 15th 1886.

He wrote in the Journal of Horticulture for 25 years, it was the rival magazine of Woodbury and the Cottage Gardener. It is said that Mr Woodbury found him a formidable opponent.

.John Lowe introduced Egyptian Bees Apis fasciata into Scotland however they did not catch on due to their bad temperament.

Another beekeeper of that time was -

William Thomson

Fig 4.1 Drawing of William Thomson

He was known as "A Lanarkshire Beekeeper" and for the Lanarkshire Hive he designed as well as the support he gave to the Stewarton Hive through his articles particularly those written in the early British Bee Journal.

William Thomson was born in 1832 in Auchinraith, Blantyre and stayed in Blantyre all his life, he died November 1898 in Lint Butts, High Blantyre, Lanarkshire. (Apparently, he went to the same school in Blantyre as David Livingstone but as Livingstone was born in 1813 it would be approximately 20 years later that Thomson attended)

He was a joiner by trade and kept bees as a hobby. However, keeping bees was to become more than a hobby.

He wrote in the Journal of Horticulture, calling himself "A Lanarkshire Beekeeper". Following the lead of Thomas White Woodbury who wrote under the pen name of " A Devonshire Beekeeper".

A Brief History of Scottish Beekeeping and Beekeepers

Fig 4.2 Photo of Thomas Woodbury

Woodbury felt it was important that readers knew where the writer / beekeeper came from, as this allowed for the seasonal differences of different areas to be taken into consideration, Woodbury also did not want to promote his own name.

Thomas White Woodbury 1818-1870 moved from London to Devon at the age of 14 when his father became the joint owner of the Exeter and Plymouth Gazette.

It was Woodbury's family that owned the Cottage Gardener magazine, Woodbury wrote articles for the Journal of Horticulture which was established in 1861 as well as occasionally writing in the Times Newspaper. Woodbury will be remembered for his hive the first moveable bar framed hive in Britain, his frame being adopted by the BBKA as the British Standard frame and the hive he designed leading to the National hive of today. He introduced Italian bees into Britain in 1859.

Thomson also wrote in the early British Bee Journal. The British Bee Journal was the idea/ dream of Charles Nash Abbott.

Abbott wanted to produce a magazine solely on beekeeping and was published weekly. Abbott also designed the 16 x 10 inch frame, that was to be later adopted by Simmins in his hive and then later in the Commercial Hive, which is still in use today.

Thomson wrote in the first magazine of *The British Bee Journal*, May 1873 writing about the Stewarton Hive as the most productive hive and producing the finest quality of honey.

Fig 4.3 Photo of Charles Nash Abbott

The Journal created by Charles Abbot in 1873 increased interest in Beekeeping throughout Britain and inspired the founding of many Beekeepers' Associations throughout Britain, including The British Beekeepers' Association 1874 and The Caledonian Apiarian and Entomological Society in Scotland in 1874. It is therefore, not surprising that he was and is called by some the Father of British Modern Beekeeping. In *the British Bee and Beekeeper's adviser* Journal Volume 1 Number 10 a proposal was made for an Apicultural Exhibition at Crystal Palace in September 1874 which required a request for funds of £100 to support the prizes of the Exhibition. William Thomson subscribed ten shillings and six pence and another Scot, John McPhedran (A Renfrewshire Beekeeper) subscribed Two pounds and two shillings.

The show at Crystal Palace was a great success although neither Thomson or McPhedran attended personally. Many of the prizes were won by Scots. Six scots won first prize for class 14 - a super of honey over 14 lbs and under 20lbs. W. Sword, J Anderson, R Graham, R Anderson, D Anderson, and A McCrone, Classes 15 and 20 were also won by J Anderson, honey produced in Stewarton Hives as well as winning a 1st prize for Class 37 - The Renfrewshire Queen Cage designed by McPhedran, made by Rowats of Glasgow and shown on behalf of McPhedran by Anderson of Dalry.

Thomson wax sheet guides were used to produce the beautiful straight combs of honey in many of the winning Stewarton supers exhibited. William Thomson also exhibited his Lanarkshire Hive at Crystal Palace but through a misunderstanding and non-compliance of the rules was not able to enter it in Class 3 of the show.

He exhibited hives and wax at many shows, winning many firsts in the process.

Thomson was credited for designing the first honey-press to extract Heather honey from honey comb with the press being exhibited at The Edinburgh show in 1884.

The Lanarkshire Hive.

It measured $17\frac{1}{2}$ inches from side to side, $16\frac{1}{2}$ from front to rear had a depth of $9\frac{1}{2}$ inches and contained 11 frames and 1 dummy board. Like the Stewarton Hive it had slides which fitted into the frames' top bars and so no crown board was necessary.

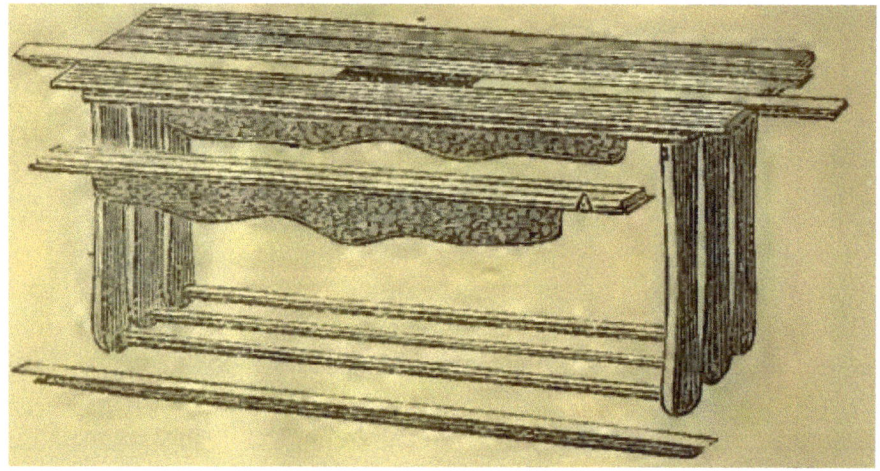

Fig 4.4 Drawings of Frames and slides of a Lanarkshire Hive

The frames ran from front to back of the hive. To help fit wax sheets the frames were fitted with a false bar A which allowed the wax sheets to be easily fitted and kept them in place.

(The False Bar had a depth of 1 inch and width of $5/16^{th}$ Inch with a saw kerf down its centre from one end to one inch off the other. When the kerf was opened with a pencil it allowed the wax sheet to be inserted. Two nails were then put through the bar to secure the wax sheet extremely securely to the frame.)

Thomson also designed other beekeeping appliances for example a cover board/ queen excluder for new frame bar hives as well as observation hives.

At one of the Honey Shows /Exhibitions, The Highland and Agricultural Society of Scotland offered a prize for the best essay on Apiculture and under the pseudonym of Pan, William Thomson won it. It was published in 1883 -*A Prize Essay on Bees* and was 84 pages long. It was the first publication in Scotland on beekeeping in frame hives and many Scottish Beekeepers moved from keeping Skeps to adopting Frame hives after reading his essay.

In 1862 Thomson was forwarded by George Neighbour and Son a wax sheet made from plates from Germany and from the sheet, Thomson ingeniously perfected a machine that produced the first wax sheets made in Britain. He wrote how he did it in Vol 2 p171 of the *British Bee Journal*. Thomson supplied his friends with wax sheets some of these friends winning prizes at the Crystal Palace Show in 1874.

When he died he was seen as a Beekeeper of the old school and that his methods were out of date however his work and contribution to the development of beekeeping in Britain as it moved from the old to modern methods will be remembered in the British History of beekeeping.

(National Beekeeping Association/Society)

After the Crystal Palace Honey show of 1874, William Thomson, J Wilkie of Gourock and others approached Col. Robert Bennett of Alloway Park and who had Offices at 50, Gordon Street, Glasgow (Robert Bennett was Royal Decorator to The Queen, Queen Victoria) about forming a Beekeepers' Society/ Association. A meeting was held in Glasgow on 28th October 1874 and the Caledonian Apiarian and Entomological Society (CAES) was set up with Sir James Watson, The Lord Provost of Glasgow as the President, Robert Bennett as Vice President, William Thomson as Secretary and Frank Gibb Dougall as Treasurer with many of the Scottish Exhibitors from Crystal Palace joining.

Ebenezer McNally became a member of the committee of the Society and he also founded and was president in 1884 of The Rutherglen Horticultural Apiarian Association. He was father of William McNally who was to become a prominent Beekeeper and commercial beekeeper in the late 1800s and early 1900s.

The Annual subscription for the new Caledonian Apiarian Society was 2 shillings and 6 pence. William Thomson was Secretary for approximately 6 years. It was seen as an opportunity for Scottish Beekeepers who could not or did not want to show /exhibit their honey further South to show in Scotland. The CAES initially was linked with the Glasgow and West of Scotland Horticultural Society and their show and then to The Highland and Agricultural Society of Scotland - many successful Exhibitions and Shows were held. However due to work commitments Robert Bennett was not able to devote enough time to the Society, to organize and fund shows and at that time was approached by Thomas Gibson Carmichael about the Society. Bennett arranged a meeting on 8th April 1891 of members of the Caledonian Apiarian and Entomological Society and others interested in beekeeping in Scotland to discuss the reorganization of the CAES or to start afresh with a new Society / Association. After discussions where it became apparent that there was still a desire for a beekeeping society, it was agreed to start anew and the 1st Scottish Beekeepers Association was formed. Sir Thomas Gibson Carmichael became secretary and spared no time or money to make the Association a success, however when he succeeded William Gladstone as MP for Midlothian, his parliamentary work prevented him committing time to the Association. He resigned as secretary of the SBA in 1895. Rev. Robert MacClelland of Inchinnan tried to keep the Association going but stepped down as secretary after a year with the Association petering out, so that by 1897 it no longer existed (Carmichael was to become Governor of Victoria and then The Indian Provinces (Bengal). In 1917 he lost his art treasure when a P&O Liner was sunk off the English Channel. The treasure was salvaged and was exhibited in Dundee in 1988.

On returning from India he lived in London but had been planning to return to Scotland and to start beekeeping again. Unfortunately, he died before this could happen. He died in 1926 at the age of 66 and by then had become Lord Carmichael.)

When beekeeping in Scotland, Lord Carmichael had tried to get his friends and neighbours to become interested in bees and beekeeping. To do so he arranged a garden party, telling his guests the benefits of keeping bees particularly to flowers and producing fruit as well as honey. He told his guests that bees were as harmless as flies and that no one should be afraid of being stung.

He invited his guests to approach the hives in his apiary and he proceeded to open a hive without protective clothing. Just as he was going through the hive a thunder clap could be heard. As Lord Carmichael lifted a frame he was stung on the head and a lady near-by started to claw at her head as a bee got stuck in her hair. The bees were angry and so most of the guests retreated quickly to the big house and the hive was quickly closed by Lord Carmichael. There were no more demonstrations arranged by Lord Carmichael - a case of once stung twice shy.

Another prominent beekeeper of that time was **William Raitt**.

Fig 5.1 Drawing of William Raitt

William Raitt was considered by his peers to be one of the finest beekeepers in Britain of his era. He was born June 20th 1839 in Newport, Fife and died suddenly, January 8th 1889, at the age of 49. He was a widower (He had been married twice) and left 7 children, 6 boys and 1 girl.

His father was a sailor. He lost his father who died in an accident, when he fell from the mast-head of his ship and died in Leith Hospital from his injuries. William was 6 at the time and was the eldest of 2 girls and 2 boys. He was clever at school where he progressed by becoming a pupil teacher, then onto Edinburgh Training College on a scholarship where he was then elected as an undergraduate of London University. He became a teacher first at Johnshaven then Hillhead in Glasgow then on to Nairn and then finally to Liff near Dundee. After 26 years of teaching (1878) bee keeping had become so profitable for William, he bought four acres of reclaimed moorland at Blairgowrie which he called Bee Croft to devote his whole time to bee-farming and so, became a commercial beekeeper.

William Started beekeeping while he was in Nairn in his early thirties after seeing / watching the bees of a friend and previous fellow student. His bees did well and when he moved to Liff he increased the number of the colonies he kept. He demonstrated and taught others on how to keep bees and was a great supporter of bar frame hives.

In 1875 William Raitt was approached by Robert Bennett and the Caledonian Apiarian and Entomological Society (CAES) to start up a Beekeeping Society/Association in the East of Scotland since the CAES had no representation for the East of Scotland.

As the International Horticultural Association was holding a meeting Bennett was hoping to hold a Honey Show in conjunction with it, which would require the support of local beekeepers and a new Society Association would be the ideal way to do this. Raitt consulted with beekeepers in the area with regards setting up an Association/Society and the running of a possible Honey Show. The result was The East of Scotland Beekeepers' Association being founded in 1876. The Earl of Airlie, the Bishop of Brechin and Miss Stirling Graham of Duntrune became its Patrons. Mr Henry Lorimer was President, and William Hay and John Stewart were the Vice-Presidents.

Fig 5.2 Photo of Robert Steele

Robert Steele was on the committee and in 1877 after the Honey Show (by that time the membership of the Association had risen to 150) would go on to set up the Bee Appliance firm that would later become Steele and Brodie of Wormit.

William Raitt was the Secretary and Treasurer and the main driving force of the new Association. He would go on to be the Associations president.

William Raitt realised the advantages in using wax foundation sheets to draw out straight honey comb, however was not satisfied by the wax sheets available at the time. In May 1877 he purchased a wax foundation machine (press), the first person in Britain to do so, from America and started to make sheets of wax available for beekeepers to buy. As it states in his obituary it was "the start of a new industry". Those South of the border took more time to realise the advantages of using wax foundation and therefore it took longer for those in the South to adopt the use of wax foundation. He advertised his super foundation as the best in the world. At the time of his death in 1889 he had 4 tons of wax in hand ready to produce foundation for the coming season.

To extract Heather Honey from the comb William Raitt designed a honey-press to do this and he sold it in his Price List as the Raitt Honey Press. It was seen to be superior to the first ever honey press - designed by William Thomson. It worked with a horizontal screw and so pushed the honey out in a side way manner without having the wax mashed up and driving the honey through it like Thomson's press. *BBJ Vol18 1890 p235.*

He also successfully grew strawberries.

It was Raitt who said " the best winter packing for bees is bees".

He wrote in many of the beekeeping magazines at that time and was sometimes editor of the Bee Section (Department) of *The Journal of Horticulture* and in 1882 wrote a series of articles for the *Weekly News* (of Dundee) covering all aspects of beekeeping. In 1885 along with William Broughton Carr became joint editor of the *Beekeepers' Record*. On his death the Committee of the BBKA recorded their regret and acknowledged the invaluable service he had rendered to the progress of modern beekeeping.

His Nephew W B Monair looked after the business initially on the death of William and then it was taken over by his son David. David unfortunately died suddenly at the age of 31 in 1903. David grew the business during his time in charge and he was in the process of expanding into the production and selling of jam, based on the production of soft fruit grown at Beecroft, Blairgowrie.

Although William died young his contribution to British Beekeeping was great and his influence touched many beekeepers including D M MacDonald and Dr John Anderson.

Miss Clementina Stirling Graham of Duntrune

Miss Graham was a descendent of Graham of Claverhouse.

Miss Graham had had an interest in bees from childhood and kept them for over 70 years and was involved in all the work carried out on her bees. To help her gardener William Spalding who managed and carried out the manipulations. Miss Graham followed entirely the methods of Jonas de Gelieu. She translated the work of de Gelieu - *Conservateur des Abeilles* for her gardener and published the translation in 1829 as *The Bee Preserver*. She dedicated the book to the Highland and Agricultural Society and Sir Walter Scott presented it to the Society and Miss Graham was awarded a Silver Medal in acknowledgement for her work. It is interesting to note that the hive used by Gilieu was a hive adapted from La Ruche Ecossaise or the Scotch Hive.

Her bees were kept in small skeps made of bent grass and she obtained her honey using straw supers.

In 1867 she sent her gardener William Spalding to check out a Woodbury Frame hive which was used by Henry Lorimer who was later to become the President of the East of Scotland Beekeepers' Association. Spalding spoke highly of the hive and Miss Graham visited Lorimer's apiary. William Spalding died not long after his visit and before Miss Graham's bees could be transferred from skeps to the wooden frame hives. When the transfer of the bees did occur in 1876 it was William Raitt who transferred the bees. Miss Graham sat very close to the bees so she could oversee the operation and every manipulation was closely watched and an explanation by Raitt was given for the action. and so, at the age of 94 Miss Graham saw a queen honey-bee for the first time. Because the transfer had been such a great success she arranged with William Raitt to remove bees from under the roof of a local Mansion. When he got to the Mansion he found a number of ladies and gentlemen had been invited by Miss Graham to see the operation take place. The removal of the bees and honey went very well, much to the delight of Miss Graham.

Miss Graham was a character and a prankster/hoaxer much to the amusement of her friends. On one occasion, disguised, she played the part of a Mrs Arbuthnott at a dinner sitting next to Sir Walter Scott. Mrs Arbuthnott spoke about her son finding a fossilised wig in a slate quarry. Scott later said he had not been fooled and had played along with the act for the sake of other guests. Others were not so lucky, being caught out by the characters and range of entertaining situations she played out. Her lawyer Francis Jeffrey was caught out.

One evening just before dinner an old Lady called Mrs Pitlyal arrived at his house asking to see him. Francis agreed and an old lady dressed in old fashioned clothes and who spoke with a broad Scottish accent was shown into the room Jeffreys was in. The lady was very talkative and her story not easy to follow and was very complicated. One of Lady Pitlyal's requests during the conversation was where could she get a set of false teeth. Francis Jeffrey gave her the name and address of two dentists nearby. After some time, partly due to the delay of her coach, the lady left. As Francis Jeffrey said goodbye politely he was taken aback by the lady. She complained of her "corny tae" as she said goodbye. Later when he was recounting the story to his wife he realised that the tiresome, strange old lady was Miss Graham. The next day he wrote her a letter returning the three guineas he had charged her as well as hoping her toe would be less troublesome in the future.

Another beekeeper of that time was -

Alexander Pettigrew 1815 to 1884,

due to the larger than usual skeps he worked with and his book *The Handy Book of Bees*.

Fig 6.1 Front cover of Pettigrew's *Handy Book of Bees*

Alexander Pettigrew was born in Carluke, Lanarkshire in 1815. His father James who was a labourer was a successful beekeeper. He saved the money he made from beekeeping to buy the Black Bull Inn in Carluke. It is said that in one year, Pettigrew's father (James) made a profit of £100, he probably made more from his side-line of beekeeping than from his actual job. The teenage Alexander and his brother managed his father's hives and this gave Alexander the initial knowledge and understanding about honey bees that he built upon to talk and write about bees later in life. James success in beekeeping was well known and this encouraged others in the area to become beekeepers and adopt his management system. Alexander in his book described his father as perhaps the greatest beekeeper Scotland has ever produced and from his book *The Handy Book of Bees* he shows he was very knowledgeable in the management of them. For example:-

In the early 1800s he was doing hive splits and artificially swarming his bees in skeps.

He also was using smoke to manage his bees - rolling a piece of corduroy and igniting one end and blowing the smoke to control the bees.

At 18, Alexander was apprenticed to gardening at Carstairs House and at 22 he moved to England. He initially worked for Lord Mansfield in his garden in Hampstead. He then moved to Wrotham Park and worked under William Thomson (not the other great Lanarkshire and Scottish Beekeeper of that time) as his foreman.

In 1844 he started to write articles for periodicals, he also moved to Yorkshire for a short period of time before moving back to London and then on to the garden of the banker, Edward Loyd at Cheetham Hill.

After moves to County Down and then Oxfordshire, where his son died from Fever, and where the lady of the house asked Pettigrew and his family to leave in fear that the Fever would be passed on to them. Alexander bought a field in Rusholme, built a house and a greenhouse and started a commercial business as a nurseryman. His business was a great success and it was during the 13 years at Rusholme that Alexander wrote his book on the management of honey bees. It was published in 1870 with the title - *The Handy Book of Bees - A Practical Treatise on Their Profitable Management*.

He sold Rusholme and moved to Sale, this time building a house and 3 vineries, growing grapes for the next 8 years.

In 1881 he once again sold his business and retired to Bowden to keep bees and sell honey.

Alexander Pettigrew died on 10th March 1884.

During his lifetime many people benefited from his experience in beekeeping. Pettigrew had huge influence on Beekeeping within Britain through his personal tutoring, communication through periodicals and his book. McNally of Glenluce sought his advice before starting up his commercial bee business in Wigton-shire.

Pettigrew inspired and influenced at least one generation of Beekeepers and is one of the giants of Scottish Beekeeping.

The Handy Book of Bees

There is much to be recommended in this book. It is easy to read and is very much about practical beekeeping. Pettigrew shows a great knowledge and understanding of honey bees and is not shy in giving his views and preferences in the management of bees, backing up his comments with observations and the comparison of results achieved under different conditions.

He writes about bees being able to communicate with one another.

"Bees have a language well understood by themselves…… and well known by beekeepers.

> The hum of contentment and the hum of trouble
>
> The hum of peace and the hum of war
>
> The hum of plenty and the buzz of starvation
>
> The hum of joy and the roar of grief
>
> The cry of pain and the music of their winter's sunshine dance
>
> The buzz of the heavy laden and the scream of suffocation."

Pettigrew made the wonderful statement -" In beekeeping there is no profitable return for foolish and unnecessary expenses". Pettigrew considered bee-houses a hindrance to good bee management and an unnecessary expense.

He supported the use of larger hives as they were more productive and more profitable.

"In beekeeping, good luck attends to those who use hives large enough to hold many bees."

"The adoption of large hives…. Would, in process of time, revolutionise beekeeping throughout the country."

"Large hives well managed are incomparably better than small ones."

"Good Management without Large hives will not end in good result." Large hives being the foundation of success and good management the superstructure.

40 to 50 years later Dr John Anderson was saying something similar.

Pettigrew carried out a survey on hive size, honey yield and date of swarming.

He suggested that queens in large and small hives lay the same number of eggs - approximately 2,000 a day. In small hives there is only room for 500 eggs for the queen to lay per day so 1,500 eggs must be eaten or removed from the hive. He argued that if you give sufficient room to allow all laid eggs and therefore larvae to be nurtured and to emerge as adult honey bees how much bigger the colony population would be and how much more honey would be produced and potentially harvested. He suggested that to get hives that weighed 100 to 168 lbs you would need the capacity of 3 ordinary hives if not more to hold the bees and produce the same amount of honey.

Pettigrew recommended 3 sizes of skeps to beekeepers.

1. A skep 20 inches wide and 12 inches deep with a capacity of 3000 cubic inches.
2. A skep 18 inches wide and 12 inches deep with a capacity of 2700 cubic inches
3. A skep 15/16 inches wide and 12 inches deep with a capacity of approximately 2000 cubic inches.

Because the widths were standard sizes, ekes could be used to increase the size of the hive by placing the eke under the hive and no alteration or adjustment was required to fit the hive.

Pettigrew stated that in fine weather a 20 inch skep would gather between 4 to 10 lbs of honey per day.

The 18 inch skep would gather between 3 and 7 lbs per day and the 16 inch between 2 to 4 lbs per day.

He preferred skeps with a flat crown rather than with a conical shape. He suggested that beekeepers should use 16 inch skeps and move up the sizes in following years.

Only if straw skeps could not be sourced should wooden boxes be considered.

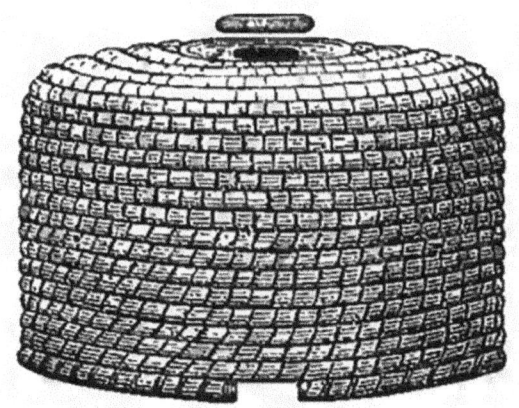

Fig 6.2 Drawing of Pettigrew skep

He saw the only positive aspect of wooden hives to be that of durability. The downside was condensation created by the bees causing the wax combs to rot making the hive useless.

He did not approve of bar framed hives either, saying they were inconvenient, clumsy and expensive and would soon go into disuse. He suggested that honey comb could be removed quicker from a skep than the time it took to remove a lid or the top of a bar framed hive. Later in life he changed his mind on moveable bar frames.

Pettigrew went onto describe how he successfully and profitably managed honey bees.

It surprised me on how similar the management of bees in the 1800s is to that of the 2020s.

Pettigrew found in skeps, bees would built comb from back to front not side to side. He describes putting in guide combs so bees would draw their comb out straight. Strips of old, clean comb 2inches wide and 2 inches deep fastened to wooden plant labels with melted wax using a hot poker. The wooden label then being nailed to the crown of the skep. Pettigrew does not give the spacing between guides or even the number per hive however I am sure it would have respected the bee space as there is no talk of issues with burr comb. Using cross sticks to support the comb (usually 5 to 6 per hive(and how bees left little holes around the cross stick which the bees would use as short cuts. Today, Roger Patterson states this is one of the benefits when using the punch cell method when rearing queens.

Pettigrew also supported a broad flight board which allowed the bees to freely leave the hive and fly off. He thought narrow flight boards caused congestion leading to a crowd and crush of bees, inhibiting ("with a lack of freedom to") the bees wanting to leave the hive.

He used covers to protect the hives – either straw hackles or felt.

He also proposed that no hive should be kept for more than 2 years – "old comb is objectionable for many reasons and ugly to look at."

The book covers many aspects of beekeeping – the honey bees' natural history, feeding , swarming, over wintering, honey production and much more. I had not known about bees producing bee ropes for the bees to ascend and descend from the comb to the entrance and how he used the thickness of the rope(s) as a guide to the readiness of a hive to swarm.

Pettigrew preferred a swarming method of bee management rather than a non swarming method.

So rather than wait for a colony to swarm naturally and have a period of around 3 weeks when no eggs would be produced. He carried out an artificial swarm by drumming bees including the queen into another skep, leaving sufficient bees to cover and look after the brood in the original hive. If the queen was not drummed into the new hive the bees would all return to the original hive and you would start again.

Pettigrew's system meant he only kept young queens in his hives killing the queens of the first swarms (earlier queens) when he united his spare swarm/skeps to his hives in the autumn with the younger queens he had produced and thus prevented natural swarming.

Pettigrew's book shows us that beekeepers who kept bees in skeps were both skilful and knowledgeable in their beekeeping and more sophisticated than just getting bees into a skep and leaving the bees to get on with it.

This book is still worth reading and there is a lot of good advice to be found.

William McNally - A Wigtown Bee Farmer

Fig 7.1 Photo of Young William McNally

William McNally was born on 17th May 1855 in Wigtownshire

He was apprenticed as a joiner working in Glasgow and Kilmarnock for 2 years between 1874 and 1876.

Spent most of his life in Glenluce - he had 6 brothers and 1 sister - had collected birds' eggs in his youth something that would be frowned upon now and is not allowed by law as well as not being seen as being an acceptable hobby, today.

He and his brother John started beekeeping in 1876 when he got a swarm of bees when on a fishing expedition.

In 1878 during a visit to friends in Manchester he was introduced to Alexander Pettigrew. He asked Pettigrew if he thought anyone could make a living from keeping bees. Pettigrew told William that he believed that it could be achieved if you had a good district for bee forage and that you had a nest egg to fall back on sufficient to cover 2 bad seasons. William McNally was therefore inspired to keep bees in earnest and to build up his bee business.

He got married in 1879 and by this time had 6 colonies, 1879 was a poor bee year and he lost all but 2 of his hives. He gradually increased his number and by 1892 had the largest number of colonies in Scotland and after this date rarely had less than 100 colonies.

In 1884 along with his brother Richard he exhibited at Edinburgh. He won most of the prizes including The Caledonian and Apiarian Society and Highland and Agricultural Society's Silver Medal for best display of honey-comb.

He sold sections, clover honey and other types of honey throughout Britain, sending it to Edinburgh, Glasgow and London by train.

In 1887 he had 153 hives and his bees produced over 3 tons of honey and he had it sold by mid-October of that same year. He continued to exhibit his honey at shows winning the silver medal at the Highland and Agricultural Society Shows for the best and largest display of honey and honey-comb five years in a row. His method of beekeeping was described by Dr John Anderson as "Let Alone Beekeeping".

He used three brood boxes/chambers with 10 frames in each.

He only manipulated his colonies 3 times in the season and did not feed his bees in the Spring or the Autumn. He left the bee colony approximately 40lb of honey stores to survive the winter period. He said that less than 1% of his hives swarmed a year and his average honey surplus per hive was 40lbs.

His first examination of the hive was around mid-June when extra room was required by the bees on their single brood box which they had been left on when preparing for the winter. The second brood box /chamber was placed under the first. Later a third brood box was added when the two brood boxes were crowded with bees. At the close of the season the surplus honey was removed and a single brood box with bees and stores were left till June the following year. McNally did not use a queen excluder. The honey was removed by pressing the honey comb and he renewed his foundation on a yearly basis.

In 1887 he wrote an article in vol. 15 of the *British Bee Journal* on Starting a Bee Farm, he echoed Pettigrew giving the requirements to become a Bee Farmer as : A Good District, Black Bees for Honey production and at least 100 hives.

In 1902 he took up growing fruit and built up a large local business selling his fresh produce which he sold at a premium since it was very fresh and clean.

He was an advocate for legislation regarding Foul Brood and supported the BBKA in its efforts to achieve this and collated information regarding the spread of Foul Brood in Scotland. Foul Brood had not been a problem when McNally had started as a beekeeper, was becoming a big issue in Scotland in the early 1900s as bees were being brought in from further South.

He wrote in the *British Bee Journal and the Bee-keepers' Record* periodically, giving practical advice, he was very much appreciated and he was held in high esteem by the beekeepers of his age.

Fig. 7.2. Photo of William McNally later in life.

He died on November 30th 1912 at the age of 57.

Second Scottish Beekeepers' Association

In 1912 The second Scottish Beekeepers Association (SBA) was formed.

George Avery was the Beekeeping Advisor of the East of Scotland College of Agriculture and it was on his suggestion that the second Scottish Beekeepers Association was launched. A meeting of beekeepers was held by the permission of the Governors of Edinburgh and East of Scotland College of Agriculture on the 27th April 1912 with the object to consider what steps should be taken with regards the formation of a Beekeepers' Association for Scotland. There was a good turn out and it was unanimously agreed to proceed with the formation of a Scottish Beekeepers Association. A committee was appointed to arrange an inaugural meeting to be held at the College of Agriculture in Edinburgh. On the 25 May 1912 the meeting was held. The Rev J W Blake who had been a member of the first SBA presided and it was agreed that the new Association be setup with the drafting of rules and regulations .

The next meeting was on the 14th of August 1912 prompted by the Isle of Wight Disease and its threat to beekeeping in Scotland. Dr T Duncan Newbigging was appointed the Chairman of the Council of the SBA and G Avery the Secretary. It was agreed there was a need for quickness on the part of the authorities in dealing with suspected disease cases.

George W Avery

Fig 7.3 Photo of George Avery

George Avery died on the 2nd of February 1945 in his 78th year at his home, The Schoolhouse Enterkinfoot, Thornhill, Dumfriesshire. He was a farmer's son, born in the village of Holborn, Northumberland, initially educated at the village school before going to Stewart's College in Edinburgh. His father had been a beekeeper and as a child he had watched his father, work his bees. However, as a boy, George through the close friendship of a lady beekeeper who was a near neighbour, he began to have a keen interest in bees. The lady supplied George with literature on bees which made him more interested in bees and wanting to know more about them and to know more about bees at an advanced level.

He had an apiary at Head's Nook when he lived in Cumbria (Cumberland). He helped set up a bee association in the Cumberland area being appointed a touring expert and later the secretary of the Association. In 1911 he was appointed to the East of Scotland College of Agriculture and in 1912 founded the SBA and became its first Secretary. Due to work pressures he relinquished the role. He retired from his position at the East of Scotland College of Agriculture in March 1933 and spent his time gardening, keeping bees and writing bee articles for beekeeping magazines including the *Scottish Beekeeper*.

Dr Thomas Duncan Newbigging
(Physician) born 1873

Lived at Kirkton of Crawford in Lanarkshire, started beekeeping in 1882. He kept between 40 to 120 colonies. He won many prizes at the Highland Show made contributions to the *British Bee Journal*, *BeeWorld* and *Gleanings*.

He designed the TDN Hive and the TDN swarm catcher, had a good collection of bee books and enjoyed curling.

Dr Newbigging, who presided at the SBA meeting in August 1912, brought forward the motion that it should be recommended to the various Agricultural Colleges and the Board of Agriculture for Scotland by the Scottish Beekeepers' Association that effective steps be put in place immediately to deal with bee disease through the appropriate experts. The motion was seconded by the Rev. R. McClelland of Inchinnan and was carried unanimously by those attending.

In 1913 concern was voiced on the delay of action in dealing with bee diseases and the impact it was having in the loss of bees in Scotland.

At the meeting on 22nd February 1913 of the Council of the SBA it was J W Moir who chaired the meeting.

The Association had a hard struggle in its early years, particularly between 1914 and 1918 during WW1. Things had looked bleak for the SBA in 1913. The number of members increased slowly with only Midlothian Beekeepers' Association affiliating with the SBA. A series of meetings were organised to consider the way forward. On the 10th October 1914 under the heading of Conference of Scottish Beekeepers' on National Federation a meeting chaired by Mr D M MacDonald. The rules revised at the Newburgh meeting were debated and amended and then sent to the Beekeeping Associations in Scotland. (It had been members of the East of Scotland Beekeepers' Association who had suggested that the SBA had to change if it was going to survive and they suggested that a new constitution was required. William Munro and Charles Nicol of ESBA (East of Scotland Beekeepers' Association) wrote the new constitution /rules.) The new constitution was adopted and Nicol took on the role of Secretary.

At the meeting in February 1915 the changes were agreed on the basis of a federation rather than affiliation. The membership number increased by six times, the number of members rose to 1,400 by the 1917 AGM and continued to grow.

Fig 7.4 Photo of William Munro

It was Munro who set up the SBA Examination system and it was William Munro who came up with the titles of Beemaster and Expert Certificates and it was he who ensured that the Beemaster exam was wholly oral and practical - so that beekeepers who would never have considered sitting a written exam could take it. To ensure consistency he produced the leaflet No 16 to allow the examiner to mark consistently the examinee as to whether or not they should be awarded the certificate.

I believe it is the form or one very like it that is used today.

The SBA owe much to the ESBA and to Nicol and Munro for its survival over the early years of its existence.

By 1919 the SBA had around 2,500 member and in a survey from 1922, a register of beekeepers in Scotland was compiled with 8,971 beekeepers and 31,281 Hives which was thought to be lower than the actual number of beekeepers and colonies that existed.

Richard Whyte 1860 -1926

Richard Whyte died tragically in a fire at his home in Pinwherry, which is not far from Glenluce, in February 1926. He managed to save one of his twin sons from the fire however while trying to save the other, both he and his son died.

He had left school to work in his father's business which weaved metal wire. He took over the business when his father died. During the first world war he was involved in making / manufacturing wire products to combat the threat of enemy submarines. His business flourished during his time in charge. After WW1 he sold the business and retired. He was then able to pursue his love of bees and beekeeping on a much larger scale, setting up a bee farm first in Cumbernauld before moving to Ayrshire and Drumspillan House, Pinwherry in 1920 where the climate was milder and therefore better for beekeeping.

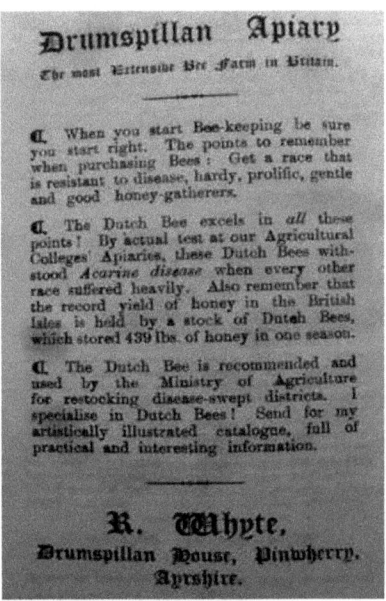

Fig 8.1 Advert Drumspillan apiary

He had started beekeeping in the late 1800's initially with 2 colonies of bees in Blantyre Hives - an early form of Bar Frame Hive. (I wonder if this was a hive made by William Thomson who lived in Blantyre at this time.)

He designed and produced many appliances including : a wire excluder; nucleus cage; Queen Mating Box and Queen Introduction and Nucleus Cage.

Fig 8.2 Whyte Queen Mating Box closed

Fig 8.3 Whyte Queen Mating Box Open

Fig 8.4 Whyte Queen Introducing Frame

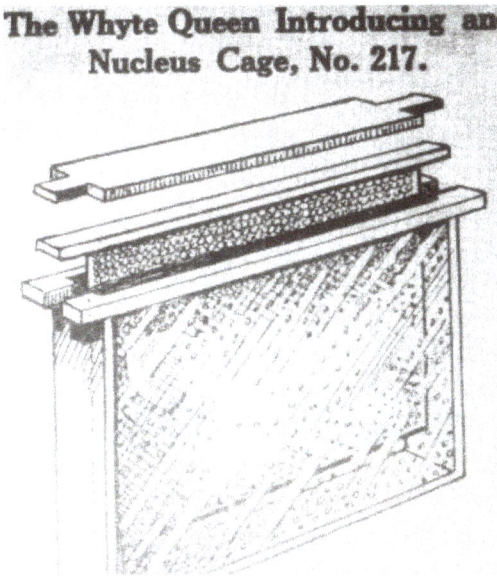

Fig. 8.5 Advert for Whyte introducing cage

He was the first president of Glasgow and District Beekeepers' Association

Richard was a keen reader and through bee books was able to build a great knowledge on bees and beekeeping. He had one of the best libraries of bee books in the country and had arranged for it to be transferred to the Apis Club (later to become IBRA) of which he had been an "originator", founder member and benefactor. Unfortunately, the library was totally destroyed in the fire with the loss of many rare beekeeping books.

The Apis Club and its' magazine The Bee World were very dear to him and he had been in the process of putting plans in place to ensure their financial viability and long term success.

He wrote articles on beekeeping and spoke at association meetings. In 1921 he spoke and wrote articles against the proposed Bee Disease Bill. Dr Anderson echoed Richard Whyte's view in his *Scottish Beekeeper* lead article June 1925 (It was considered during the early 1900s that disease threatened the very existence of beekeeping and that steps and representation was required for effective action to be taken to deal with bee diseases as well as involve the experts of the day. This action, in no small part, led to the formation of the SBA in 1912 and local Associations who could then make representations to the various colleges as well as politicians).

In 1922 at a meeting of Kilmarnock and District Beekeepers' Association Richard Whyte did a talk and presented Silver Medals from the Apis Club for Honey Exhibits at the Ayrshire Agricultural Association Show that year. At that meeting Joseph Tinsley was presented with a timepiece for the series of lectures he did for the Kilmarnock Association. At the interval Major David Yuille sang a song and the vote of thanks was given by John J Walker of Kilmaurs.

I wonder if in 1918 it was at a series of Lectures carried out by Joseph Tinsley that Glasgow and District Beekeepers' Association was set up and this is why Joseph Tinsley said that he was in good part responsible for GDBKA's "inauguration".

(Glasgow and District Beekeepers' Association was originally formed by the members of a class held at the Agricultural College, with 53 people joining with Richard Whyte as President and Peter Bebbington as Secretary (I believe John J Walker was also a member). At the first Annual Meeting on April 28th 1919 the total number of members was 97 with only 2 members resigning from 1918. Major Henderson Member of Parliament for Tradeston was made Honorary President of the Association in 1919. At the meeting it was proposed that there should be monthly meetings as well as a series of beekeeping lectures.

The association also affiliated/ linked with the SBA at that time.

Richard Whyte set up a process to buy and distribute Dutch Bees throughout Scotland. It was on his visits to Holland that he became good friends with Joseph Tinsley. (Tinsley also had to visit Holland to buy bees to restock the West of Scotland Agricultural College.)

Whyte imported hundreds of stocks of bees into Scotland and the losses in transit were high. He also gifted colonies of bees from his own stock when he was made aware of a beekeeper who had lost all his / her stock from Isle of Wight disease. He was also involved in the scheme for small holders to start up in beekeeping. He funded the same amount of money as the Board of Agriculture for Scotland so that the smallholders could purchase their bees at half the price they would have normally. More than one hundred smallholders benefitted from his generosity in the first year of the scheme alone.

He also judged at honey shows - in 1924 he was the honey judge at the prestigious Ayrshire Agricultural Association in Kilmarnock that year.

Fig 8.6 Richard Whyte Judging at the Ayrshire Agricultural Association Show 1924

Associations visited his apiaries where Richard and his family gave them a warm welcome and great hospitality. One of those visits was described in the First *Scottish Beekeeper* July 1924 page 13 when The Ayr Association visited and he showed a number of his hives and appliances and gave advice from his vast beekeeping experience including the need to work strong colonies.

Richard Whyte was a great Scottish Beekeeper. He made a great contribution to beekeeping in Britain through his appliances , also by being a great benefactor to Scottish Beekeepers as well as the Apis Club. It is unfortunate that he died so tragically and that beekeeping lost such a great spokes-person, benefactor and supporter of beekeeping in Britain.

The area around Pinwherry in Ayrshire and Galloway is ideal for keeping bees in large numbers. In this area in the early 1930s Hugh Howatson Senior and his son Percy kept bees Commercially. After the death of Percy Howatson in the late 1940s, Hugh Howatson Junior took over at Garlieston and kept around 400 colonies for around 30 years in that area. Selling to the supermarkets of the time as well as selling honey and sending it to the then President of the United States of America, Dwight D Eisenhower who had tasted the honey on one of his visits to Culzean Castle where he had an apartment.

Until recently John Mellis has been a commercial beekeeper in the Dumfries and Galloway area.

John Wilson Moir and the Moir Library of the Scottish Beekeepers' Association (SBA).

Fig 9.1 Photo of John Moir

John Moir and his younger brother Fred, from Edinburgh, in 1878 followed in the footsteps of David Livingstone to East Central Africa to carry on his work in the development in trade and agricultural development. At that time the Free Church of Scotland's Livingstone's Mission formed a Company (based in Scotland) The African Lakes Company (the company existed up until 2007 when it went into liquidation). The objective of the company was to support the abolition of the Slave trade through Christianity and legitimate commerce. The biggest shareholders at that time were the Moir brothers and both were appointed joint manager. During their period of steward-ship they successfully built up an export trade. The company introduced new crops such as rubber, tea, coffee and tobacco. The first coffee plants coming from Edinburgh's Royal Botanical Gardens.

Not everything went smoothly, in 1887 conflict broke out between the African Lake Company and the Arab Slavers whose businesses they had compromised. In the fighting John was wounded when shot through his thigh and Fred who was shot through his right arm which was shattered below the elbow. After 2 years and 3 months the African Lake Company defeated the Arab slavers.

In 1890 John settled down as a planter in the Shire Highlands and it was then when he took up beekeeping to aid with the pollination of the plants he sowed. This was to be the start of a hobby and an interest in honey bees that would last the rest of his life.

After a serious illness he returned to Scotland in 1900 and once recovered took up Charitable and Social work as well as keeping bees.

In 1912 he was a founder member of the reformed Scottish Beekeepers' Association and it was then he decided to seriously start to keep bee and beekeeping literature. He worked systematically to improve the depth and range of beekeeping books.

In 1916 he had collected 167 books and it was then he decided that when he died he would leave his collection to the SBA. Later he decided to allow members of the SBA to access the books during his lifetime and that he would continue to keep the collection at his house and act as librarian.

John Moir corresponded with beekeepers throughout the world e.g. Dadant in the USA. In exchange for sending duplicate books some of which were old and rare, numerous American and Canadian beekeeping books found their way to Scotland and the Moir library.

Up until 1931 he bought books for the library with his own money however due to the share collapse and the great depression his income in real terms was much reduced. In 1932 he therefore gifted the collection to the SBA who set up a process to set aside money annually for the library and in 1934 a library fund was instituted.

In 1933 John Moir wrote to Edinburgh Public Libraries (EPL) seeking advice on the binding and repair of some books and thus a relationship was started between the libraries and John Moir.

In 1935 the size of the collection of books was becoming more difficult to house and so John Moir proposed to the SBA that the collection be housed in the Public Library after his death or even earlier if required.

In 1936 all the legal documentation governing the transfer of the collection to EPL had been set up. John continued to buy books for the collection however he suffered a stroke in 1937.

In 1938 at the Empire Exhibition approximately 200 duplicate books were sold with the residue of unsold ones being sold to the American Bee Journal in early 1941.

On the 6th of October 1939 the Moir Collection was transferred to Edinburgh Public Libraries. In 1939 the Moir Library was housed in the George IV Bridge and neighbouring building until 1993, moving in March that year to Fountainbridge Public Library, Dundee Street. After flooding in 2001 some of the rarer and more valuable books (around 250) were put on loan to the National Library of Scotland.

John Moir died on the 13th of March 1940.

John Moir through his foresight created a collection of beekeeping books that is unique, containing many rare books as well as a wide range of French, German, Italian and American books and periodicals. Through this collection every member of the SBA has access to the history and knowledge of all aspects of beekeeping. No wonder it is considered to be the greatest and most prized asset of the SBA.

To make the collection more accessible there have been suggestion that parts of the collection be distributed to other Libraries throughout Scotland as well as books being scanned /digitalised and members being able to access the ebooks /magazines, an example of this, is members being able to access the archives of the Scottish Beekeeper Magazine.

Some of the books in the collection gave John Moir a great deal of pleasure when he was able to secure them. Reaumur's *Memoires pour server a l'histoire des insectes* was one of them.

Another was the book by Remnant. In 1938 he discovered Remnant's *Discourse or historie of bees* for sale. He asked Mr Savage of Edinburgh Public Libraries on advice on how much he should offer. Not long after the discussion, Savage received a postcard with a message scribbled by Moir. It said " Dear Mr Savage, I got the Remnant for £3. 10/- Hurrah! Yours rejoicing John W Moir."

When Topsell's *Historie of foure-footed beastes* was acquired by John Moir , he shouting "Minto, Minto, come and see this I've waited 15 years for this book."(Minto was Curator of Edinburgh Public Libraries)

Another treasure of the Moir is Daniel Wildman's book, *A Complete Guide to the Management of Bees* 1780.

Wildman was the first British Beekeeping Appliance Maker, had a Bee and Honey Warehouse supplying a variety of types and different designs of appliances to harvest wax and honey without killing the bees. His book gave a lot of good advice, he suggested beekeepers leave sufficient honey so that they would not starve during the winter because of owners robbing them beyond all reason.

Other treasures of the Moir include Books by the Scottish Beekeepers : Robert Maxwell's *The Practical Beemaster*, 1747; James Petrie's *The Scots Apiary* 1769 ; Howatson's *The Aparian's Manual* 1827.

The manuscript *On the Care and Knowledge of Bees* by James Playfair (1752-1812). The book was never published and had thought to have been destroyed in a fire at the printer's office only to surface later and be returned to his family.

The translation of J de Gelieu book by Miss Stirling Graham *The Bee Preserver* (It was only after the 2nd edition that she admitted to having written the translation).

Moir also had disappointments in trying to acquire books for example, the Book on Beekeeping by Thomas Hill - *Profitable Arte of Gardening* with a section of *Profitable Instruction of Perfite (Perfect) ordering of Bees*, 1st Edition 1568 became available from Maggs Brothers at a price of £31 10/- (A huge amount of money in those days). Moir was unable to secure the funds through the SBA quickly enough and lost the book to another buyer (probably American). Later he managed to acquire the 1593 edition much to his delight.

This book was the first book on bees printed in English. The details had been compiled by Hill but it is thought to be the work of George Painter - Hill was not even a beekeeper.

The depth and quality of Books in the Moir library is amazing and something all Scottish Beekeepers should appreciate and be grateful for the foresight and energies of John Moir.

Donald McIntosh MacDonald

Fig 9.2 Photo of D M MacDonald

D M MacDonald was born Grantown on Spey, Morayshire in 1853 and died at Grange near Keith in 1928.

He trained as a teacher in Edinburgh and was appointed Master at the public school at Morinsch, Ballindalloch, Banffshire around 1884. He received the Honorary degree of Fellow of the Educational Institute of Scotland in recognition to his work in education.

He bought his first bees in 1889 and was taught how to keep bees by Mr Stokes of Duthill, Strathspey, a disciple of William Raitt. His advice to new beekeepers was "to read every bee book you could lay your hands on especially Cowan's Guide". Herbert Mace wrote that it was said that MacDonald was such an avid reader that almost the entire BBKA library found its way to his home by instalments. Mr Stokes was an advocate of keeping very strong colonies which would produce a good honey yield and this was reflected in D M MacDonald's beekeeping.

He was a successful beekeeper and with the honey that he produced, beekeeping went from a hobby to one which gave him a return of between £1 to £2 per hive each season. He kept between 20 to 30 hives.

Over time he joined the staff of contributors (Unpaid) to the *British Bee Journal and The Beekeepers' Record* which he frequently wrote for (Almost weekly) and the quality/value of his practical beekeeping advice was such that his articles were reprinted in Bee Journals in America, Canada and Australia.

D M MacDonald was an advocate of Native Black Bees. He wrote "Bees have been bred too much for their face value, for colour or good looks and the result is that we have been getting softer bees... less hardy, less industrious, less gentle and less able to resist disease... This Strain (Yellow bees) has converted our fine old race into a set of mongrels... tainting the apiaries of even those beekeepers who would not have them."

Regarding over manipulation he wrote :- "More injury is worked by over than under manipulation. Pulling up a plant by the roots to see how it is growing is not good for the plant. Opening up hives to see how the works are progressing is not good for the bees. Blowing clouds of smoke into the eyes of the busy little toilers cannot tend to make them more comfortable. Pulling frames apart to see trifling nothings must interrupt the traffic, demoralise order and upset well matured plans." Manley echoed MacDonald's words in his book Honey Farming.

Although a member of the BBKA he felt it only represented the Beekeepers of London and its surrounding areas and did not represent the views and did not voice the opinions of Scottish Beekeepers and therefore the BBKA was not really a British Society /Association. Through his writing he suggested changes to the BBKA so that it was more representative but those suggested changes were never made. MacDonald felt strongly that Scottish Beekeepers should not be "left out in the cold" and was therefore a strong supporter of a Scottish Beekeepers' Association (SBA) and was a founder member of the new Scottish Beekeepers' Association in 1912.

D M McDonald was appointed Instructor of beekeeping to the North of Scotland College of Agriculture in 1913.

He was Chief examiner for BBKA Expert Certificates in the North and in 1918 had to examine Dr John Anderson for the BBKA third class certificate. After the examination started he told John Anderson he was "ashamed to ask him such elementary questions - so come and have some tea". So, the examination was finished over a cup of tea.

D M MacDonald died in May 1928, he was a founder member of the SBA, Honorary Life Member of the BBKA, Honorary Expert Bee-master and Honorary Honey Judge of the SBA.

Aberdeen Association made him a life member and Honorary Vice President.

Because of his writings not only was he well known in Britain he was also well known all over the world as one of Britain's most eminent beekeepers. Unfortunately, he never wrote a book on beekeeping - he had considered it but felt he could not improve on the books of Cowan and Digges so decided against writing one.

If you get the chance, read some of the articles that he wrote, they influenced a generation of beekeepers including Herbert Mace and ROB Manley, they are so very much worth a read.

Dr John Anderson and the Glen Hive

Fig 10.1 Photo of Dr John Anderson

In 1918 John Anderson designed the 15 frame double walled hive known as the Glen Hive. In the February 1926, *The Scottish Beekeeper* he wrote "I recommended hives take not less than 15 British Standard (BS) frames and one firm put those on the market." (Probably J & A Ogilvie of Union Street, Aberdeen).   See Advert

Fig 10.2 Advert for Glen Hive J&A Ogilvie

The benefits of the bigger hive were :

1. Better swarm control with less frequent swarms

2. Less manipulation of the hive was required

3. Sufficient space for the queen to lay and to store surplus honey, resulting in bigger honey yields and better overwintering of the bees.

4. Height of the hive with supers was reduced for large yields of honey in comparison to a single or double 10 BS hive and its supers.

Anderson was not fixed on using BS standard frames, in his article in the July 1925 *Scottish Beekeeper* stated that it should be 15 BS frames or equivalent, however, due to the variability of our Scottish climate and weather he stated that the hive should be double walled (which allowed for packing e.g. Straw) and that the roof should/would be waterproof.

Dave Cushman's website describes the Glen Hive - "Looks like some type of WBC hive which holds 15 frames per box and was believed to be intended for honey production on Scottish moorland." They were very heavy and difficult to move. I have read that they were moved like sedan chairs with stretchers to carry them. I have also heard that horse drawn carts were used to take them to the heather. I think though it is interesting that Anderson himself in one of his articles (Bigness) wrote. "Many will urge that big stocks in big hives with big stores will be difficult to lift"... (but as Dadant said "why lift them"). We never need to move those hives, and we shall not lift even the supers,... When extracting, we shall take out the super combs one by one, shaking or brushing off the bees." So, I wonder if the transitory use of Glen hives was more inventiveness by other Scottish beekeepers rather than something Anderson had considered, planned or even intended doing.

I think it is hard enough moving a 10 BS framed single walled hive to the heather never mind a 15BS framed double walled hive.

If you are interested in seeing how a Glen hive is worked check out the guest blog by Paul and Jean on the 3 July 2020 written in the wonderful blog of the Beelistener (Ann Chilcott) or speak to Alan Riach who has worked one at the EMBA association apiary although I think it is more of an exhibition hive now rather than a working hive.

Fig 10.3 Photo of Glen Hive from *The Scottish Beekeeper* 1925

Dr Anderson spent his early life on Orkney, he graduated from Aberdeen University in Both Arts and Science and later was awarded his Ph.D.. He was a science teacher at the Nicolson Institute on Stornoway and it was there that in 1910 he designed the Nicolson Observational hive which he used for educational purposes. An article was published in the May 1911 *British Bee Journal* on how to use the Nicolson Hive.

See illustration of Nicolson Hive.

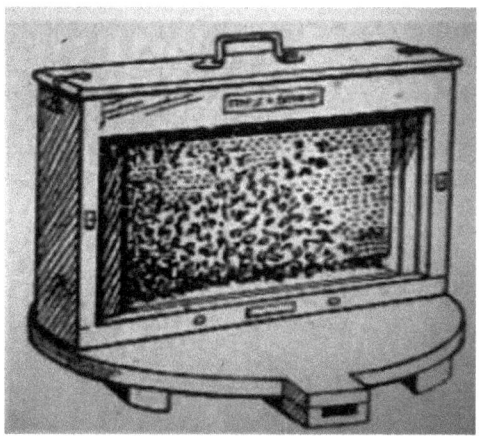

Fig 10.4 Drawing of Nicolson Observation Hive

I don't think there are many Nicolson Hives about now, the only one that I am aware of is owned by Alan Riach and is part of his great talk on Hives Through the Ages. There may be one at the Nicolson Institute where in the early 1960s one was still being used for educational purposes.

In 1915, John Anderson became the first lecturer in Beekeeping at the North of Scotland College of Agriculture. He was the person who set up the apiary at Craibstone and lectured and demonstrated all over Scotland. He wrote articles for *The Scottish Beekeeper*, other British and Foreign Beekeeping Journals. He was considered to be an expert in handling bees and was described by Mrs Shepherd as one of the best she had ever seen.

He was chairman of The Aberdeen District Beekeepers' Association which at that time had more than 1,600 members and was the largest Bee association in Britain. He was Honorary President of The Glasgow and District Beekeepers' Association and also had been President of The Apis Club. He was president of the Scottish Beekeepers' Association in 1919, writing the lead Article in the first editions of *The Scottish Beekeeper*, then becoming its editor from 1926 until his sudden death in 1939.

John Anderson believed strongly it was better to keep bigger colonies than small ones and he used *The Scottish Beekeeper* to publish his articles supporting this point of view.

However, in the first edition of *The Scottish Beekeeper* he wrote about "A Higher Standard".

In this article he puts forwards the objectives of the SBA and *The Scottish Beekeeper*. The objectives were to achieve a higher standard of knowledge of beekeeping in Scotland. He discussed how this was being achieved at that time:- through the education system and College lecturers in beekeeping. The need to be more consistent in delivering good and better honey seasons by raising beekeeping standards. He wrote of the need to increase business standards supplying consistent quantities of honey at a fair price (i.e. stop the inconsistency of low harvest and high price of honey or good harvest and low honey price, caused through supply and demand. With good hive management there should be no bad seasons) along with the need for beekeepers to highlight to the wider public the virtues and benefits of honey.

He also wrote about the need for beekeepers to have a higher standard of co-operation and camaraderie. He suggested that all beekeepers "shall work together for common ends and that we shall not tolerate petty jealousies or sectional activities. Natures gifts are diverse and we have come by different ways but we all have something to contribute to the common fund and none should withhold his contribution but give freely according to his talent". These words still resonate and I am sure many Associations as well as the SBA trustees try very hard to meet those objectives and expectations today.

He believed in natural resistance of bees to disease and argued that by treating bees for diseases - (e.g. American Foul Brood and Isle of Wight disease) only promoted the survival of unfit stock and breeding in susceptibility. He tried to breed and promote breeding from disease resistant stock, however, his work did not get sufficient support and his work came to nothing.

John Anderson did not support the thinking on Isle of Wight disease which was proposed by the Cambridge University team led by Dr Malden which believed the disease was caused by the microsporidium *Nosema apis*. Anderson had found bees with Isle of Wight disease without Nosema and was therefore sceptical of the Cambridge groups conclusions. He collaborated with Dr John Rennie and his work around the Isle of Wight disease. Anderson therefore helped in some way to identify that Acarine disease was caused by a mite - *Acarapis woodii*.

His second lead article in the August, *Scottish Beekeeper* 1924 was titled "Bigness" where he advocated the need for bigger hives to meet the needs of brood and a good queen. Dr Anderson used what he describes as simple arithmetic in his reasoning for a bigger hive. He argued that a moderately good queen could lay 3,000 eggs per day so the minimum space required by the queen over a three week period would be 63,000 cells with another 20,000 cells being required for food (no less than 10lbs of honey and space for sufficient pollen) a total of 83,000 cells - 15 BS frames (83,000/5600 =14.82 rounded up to 15). It is interesting that around 50 years later Ian Craig used similar simple arithmetic in an article in *The Scottish Beekeeper* to support his 8 + 8 British Standard frame double brood chamber management system.

When Dr Anderson died suddenly in 1939 from a heart attack the Scottish Beekeepers' Association set up their highest award in his memory.

The Dr John Anderson Memorial Award is given in recognition for work and service in the furtherance of Beekeeping in Scotland and beyond which very much supports John Anderson's work and philosophy on improving the standards of beekeeping in Scotland, through improved practical skills, better knowledge and education.

Much of the work that is carried out by the Scottish Beekeepers' Association today is still influenced by the actions and thoughts of Dr John Anderson.

Dr John Anderson was a Giant in Scottish Beekeeping during his lifetime and will be remembered forever as one of Scotland's greatest beekeepers.

Dr Edward P Jeffree and Alex S C Deans
(Beekeeping research at Aberdeen).

Many are aware of the research carried out by John Rennie, Bruce White and Elsie Harvey which showed that Acarine disease was caused by a mite, *Acarapis woodii* named after the benefactor who funded the research.

(In 1919 Rennie would become the first President of the Apis Club.)

Research in Aberdeen continued through Dr G D Morrison who set up the Beekeeping Research Department in 1947 (Guy Morrison was awarded the Dr John Anderson Memorial Award in 1951) E P Jeffree joined the department at that time.

Dr Edward Jeffree

Edward Jeffree was born in 1908 and died in 2004. Edward had a degree in Chemistry, Maths and Physics from London University. Before going to Aberdeen, he had worked at Rothamsted under Dr C G Butler.

With Dr Delia Allen and Charles Cooper, they looked at long term causes and effects in beekeeping, this included the weather and its effect on flowering times the effects on colony development as well as the impact temperature had on acarine disease.

E P Jeffree gave some of his time over to the SBA Research Committee and was convener of the committee from 1949 until August 1960 when he moved to Orpington and took up a teaching post in Hull. Dr Jeffree was an Honorary Associate of the Scottish Beekeepers Association and was awarded a D.Sc. in 1958 and during his period as Convener wrote articles under the heading of Research Notes in the *Scottish Beekeeper* magazine covering such areas as Spring Stimulation and colony development (Oct. 1950), drifting, robbing and swarming (May 1951), queen introduction (Sept. 1951), wintering, Acarine and Nosema diseases (Aug. 1951).

In the *Scottish Beekeeper* in April 1947 p 73-74, E P Jeffree wrote an article on **Dates for the Migratory Beekeeping in Aberdeenshire and Adjacent areas**. In it he describes the significant sources of nectar and pollen along with the flowering times for the North East of Scotland – if you get the chance you should read this article, although flowering times will be slightly different in different areas/ locations, as well as, through the changes to our climate/ weather. This article will give you an idea of the important nectar and pollen sources for bees in Scotland as well as a rough idea when these plants flower. Jeffree's aim in the article was to highlight two important factors in migratory beekeeping - flowers that were important nectar/ honey sources and when they flowered.

His findings on the 10 best sources for migratory beekeeping are shown in the table below.

No.	Common Name	Latin Name	First blooms	Peak blooming	Nearly over blooming
1.	Gorse/ Whin	Ulex Europa	26th March	18th April	9th June
2.	Blaeberry	Vaccinium myrtillus	1st April	4th May	5th June
3.	Sycamore	Acer pseudo-plantus	5th May	16th May	28th May
4.	Broom	Cytisus scoparius	2nd May	7th June	2nd July
5.	Raspberry	Rubus Idaeus	4th June	16th June	2nd July
6.	White Clover	Trifolium repens	16 June	13th July	19th August
7.	Bell Heather	Erica cinereal	2nd July	20th July	8th August
8.	Lime	Tilia Species	17th July	27th July	9th August
9	Willow Herb	Epilobium angusifolium	9th July	27th July	31st August
10.	Ling Heather	Calluna vulgaris	10th August	26th August	14th September

It is also interesting to note that in 1949 in a talk to the East of Scotland Beekeepers Association, E P Jeffree suggested that there was a link between swarming and flowering times. He stated that the time bees were most likely to swarm in the North East of Scotland was between 20th May and 20th July and that there was a definite clue to swarming from the lupin. He stated that when the Lupin was flowering and was fully out you needed to look out for swarming and when the flowering was over, swarming was over. I don't have any lupins in my garden so cannot comment on how true this is, however I am seriously thinking of getting some to check his theory out. Now I know where the saying "Lupin time is swarming time" came from.

(In the recent book by Aston and Bucknall, *Good Nutrition Good Bees* stated that in most cases, plants were flowering earlier than 100 years ago and they postulated earlier flowering would be beneficial to honey bees.)

Alex S C Deans

Fig 11.1 Photo of Alex Deans

On Dr John Anderson's retirement A S C Deans became Head of The Beekeeping Department of The North of Scotland College of Agriculture (Margaret (Peggy) Logan was a member of the Department at this time).

A S C Deans was also part of the Scottish Beekeepers Association's Standing Committee on Education. He was awarded the Dr John Anderson Memorial Award in 1965. He wrote three books, the first *The Bee Keepers Encyclopaedia* a book on practical beekeeping along with a beekeepers' calendar, his second book was called *Beekeeping Techniques* and finally in 1962 *Bees and Beekeeping* a beekeepers' practical handbook part of The Oliver and Boyd Quest Library.

These books are out of print but are still readily available on eBay /Northern Bee Books.

The book *Beekeeping Techniques* covers many aspects of Beekeeping from Bee Breeding, Honey production, bee disease, control and diagnosis as well as microscopy of the honey bee, pollen analysis including in honey along with honey analysis i.e. the constituents of honey. It is a book which contains a wealth of information.

11.1.1 Photo of book cover of *Bees and Beekeeping*

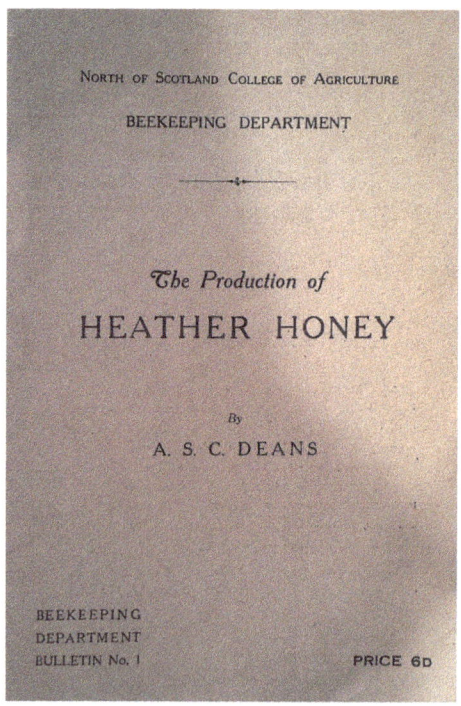

11.1.2 Cover of NSCA Bulletin No1 on The Production of Heather honey by ASC Dean

A S C Deans was acknowledged as an expert on Heather honey production and his pamphlet : *The production of Heather Honey*, A S C Deans, North of Scotland College of Agriculture, Bulletin No. 1 was considered to be the best source of information available on the subject.

It was considered that some of the publications on the subject lacked sincerity i.e. the author was not writing from personal knowledge or practical experience.

Deans through his work on the Biometry of bees tried to predict (to some success) how good a colony would be at producing honey. The method used was comparing the cubital index of the bee wing and relating this to honey production. It is interesting that others including BIBBA used similar techniques to identify sub species/ strains of honey bees after this work.

In March 1953 A S C Deans' article on Honey Sources was published in the *Scottish Beekeeper* this was the first large scale survey of honey sources to be published in the United Kingdom. Deans had found that the best way to identify the origin /source of honey or honey dew was through pollen analysis. The methods/ techniques and results obtained were published in his book *Beekeeping Techniques*.

Deans highlighted that the honeys displayed on the show benches were all different since the flower sources of nectar /honey gathered by the bees were all different.

Deans took 100 samples in 1950 of honey selected randomly from throughout the UK - from the Shetland Isles to the South of England. Deans suggested that from the results "it gave one an idea of the popularity with bees of the various nectar yielding plants" throughout Britain.

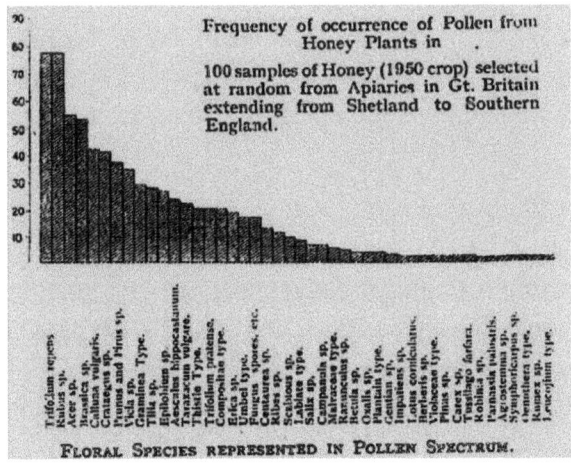

Fig 11.2 Frequency of Pollen Occurrence in the 1950 Study

Plants Represented in Pollen Spectra of Honey Samples, Season 1950.	
Latin Name.	Common Name.
Trifolium repens.	White Clover.
Rubus spp.	Raspberry and/or Blackberry Bramble.
Acer spp.	Sycamore tree.
Brassica spp.	Charlock, Mustard.
Calluna vulgaris.	Ling heather.
Crataegus spp.	Hawthorn.
Prunus/Pyrus spp.	General fruit group—Apple, Pear, Plum, Cherry.
Vicia spp.	Field bean, Vetches or tares.
Graminea spp.	Grasses—Wind-borne Pollen, indicating presence of honeydew.
Tilia spp.	Lime tree.
Epilobium spp.	Willow herb.
Aesculus hippocastanum.	Horse chestnut.
Taraxacum vulgare.	Dandelion.
Cirsium spp.	Thistle.
Trifolium pratense.	Red clover.
Compositae type.	General heading embracing coltsfoot, yarrow, sunflower.
Erica spp.	Bell heather.
Umbel type.	Hogweed or cowparsnip.
Centaurea spp.	Knapweed, cornflower, yellow star thistle.
Ribes spp.	Gooseberry, Black, Red and White Currant.
Scabiosa spp.	Field scabious, Devil's bit scabious.
Labiate type.	Catmint, Horehound, Marjoram, Sage. Salvia.
Salix spp.	Willow.
Campanula spp.	Harebell.
Malvaceae spp.	Mallow.
Ranunculus spp.	Buttercup.
Betula spp.	Birch tree.
Oxalis spp.	Wood sorrel.
Plantago spp.	Plantain.
Gentian spp.	Gentian.
Impatiens spp.	Balsam.
Lotus corniculatus.	Birdsfoot trefoil.
Berberis spp.	Barberry.
Violaceae spp.	Wild viola or pansy.
Pinus spp.	Fir tree.
Carex spp.	Sedge.
Tussilago farfara.	Coltsfoot.
Robinia spp.	False acacia.
Parnassia palustris.	Grass of Parnassus.
Agrostemma spp.	Corncockle.
Symphoricarpus spp.	Snowberry.
Oenothera spp.	Evening primrose.
Rumex spp.	Dock.
Leucojum spp.	Snowflake.

Fig 11.3 Plants represented in the 1950 Study.

Aston and Bucknall's Book *Good Nutrition Good Bees* refers to Deans work and in particular his thesis submitted to the National Diploma in Beekeeping Board regarding useful nectar yielding plants in England compared with Scotland, Ireland and Wales. He concluded that the most important plants as a source for honey were:-

White Clover (*Trifolium repens*), Prunus/ Pyrus species, Acer species, Castanea species, Tilea species, Brassica species, Ligustrum species, Vicia species, *Trifolium pratense*, Chamaenerion species, Cirsium species, Campanula species, *Calluna vulgaris*, Heracleum type plants, Tanaxacum type plants, Centaurea Species, Aesculus species, Erica species. He also concluded there were only two single source type honeys normally gathered - White Clover and Ling honey (Oil Seed Rape would be included as a single source type honey today).

Little did Deans know the importance of this work and the aim of looking at the long term causes and effects in beekeeping.

Because of Deans work, comparison studies can be carried out to look at the effect of changes including weather compared to those of the 1950's. Rex Sawyer in the 1983 *Scottish Beekeeper* p 8 in his article *Progress in Beekeeping,* acknowledged Deans work and the need to repeat this work and evaluate the changes brought by modern agriculture and climate change.

Aston and Bucknall highlighted the following surveys of honey sources in Britain by the BBKA in 1985, 1994, the Bee farmers Association in 1994 and one carried out in 2020 and published in Nature by Jones et al. https://doi.org/10.1038/s42003-020-01562-4. Using Deans work, it showed a similar correlation of honey sources over time. However, the 2020 paper/ 2017 survey showed that changes in Agricultural practice, crop use and intensification along with the spread of invasive plant species had altered nectar and pollen sources available to honey bees and with those changes in foraging plants, had led to changes in the types of floral honey available.

Joss Bartlett in his book *The Pollen Landscape* – In North Wales, out of 61 types of pollens collected in pollen traps from February to October 2019, identified 7 predominant sources of pollen – They were Hazel (*Corylus avellana*) in February. Willow (Salix species) February / March Fruit trees (Prunus species) April, Hawthorn (*Crataegus monogyna*) May, Oil Seed Rape /Sainfoin (Brassica) June/ July, Clover (*Trifolium repens*)Late July to Early September and Ivy (*Hedera helix*) October. He suggested that some of the secondary sources of pollen could be just as important due to the bees collecting them over a longer period of time than the predominant pollens e.g. dandelions (Taraxacum species) Rosebay Willow Herb (*Chamaenerion angustifolium*). Only the Hazel an important source of pollen for spring colony build up and Ivy an important late source of nectar for winter are not listed by Jeffree or Deans.

Going forward it is important that as beekeepers we are aware of the nectar and pollen sources that are available to our bees, their survival may depend upon it. Remember when nectar or pollen are not available or available in sufficient amounts to free flying bees – (from the countryside / our gardens) the bees then become dependent on the honey and pollen they have stored or the food that we give them. No food source and the bees and other species will die. It is not just honey bees we need to protect and help survive.

We can plant flowers, shrubs and trees that will give honey bees and other pollinators additional nectar and pollen sources e.g. Snowdrops, crocus, winter aconite bulbs in the autumn to flower in the spring .

Read *Plants and Beekeeping* by F N Howes which has chapters on major honey plants and plants visited by the honey bee for nectar and pollen. If you are in a Beekeepers' Association they will have this in their library

The Book *Beekeeping for Gardeners* by Richard Rickitt has a section on plants and their value to pollinators which will help in the choice of plants for the specific type of garden you want and which promotes biodiversity and sustainability.

We can observe at the hive entrance and identify the pollen being brought in using the pollen colour charts in William Kirk's book *A Colour Guide to Pollen Loads of the Honey Bee*.

Take up microscopy and identify the nectar and pollen sources of your bees – Rex Sawyer wrote two books *Pollen identification for beekeepers* and *Honey identification* which along with Margaret Anne Adams' book *Pollen Grains and Honeydew – a Guide for identifying the plant sources in honey* has all the information you will need to get started.

Follow the work of Jeffree and Deans, identify the nectar and pollen sources of your bees and make sure your bees have good nutrition and help the long-term survival conditions for many plant and animal species.

Dr Mary Delia Allen

Fig 12.1 Photo of Dr Delia Allen (Seager)

Like many other significant beekeepers who have made a notable contribution to beekeeping, Delia's memory has faded over the years.

Delia Allen was born 1929 and lived her early life in Wanstead. Her father was a lecturer in Chemistry.

She enjoyed painting, playing the cello and gardening.

She was a Quaker and made considerable contribution to church/ meetings.

In the 1950s Delia Allen became a research scientist based at the Beekeeping Research Department, North of Scotland, College of Agriculture, Marischal College, Aberdeen. She was a colleague of Dr Jeffree who also worked at the college.

Most of her research was carried out from around 1955 to 1965.

Delia Allen was awarded a Ph.D. from London University in 1958 - her thesis was entitled:- *A study of some Activities of the Honey bee*. It was in two parts.

Part 1:- *Behaviour of honey bees with particular reference to the Queen and Swarming which includes a study of the activities and ages of the Queen's attendants and investigation into the SHAKING of both Queen and workers and a description of the events proceeding the departure of swarms.*

In a talk to Glasgow and District Beekeepers' Association in 1963 Dr Allen talked about how she made her observations. Bees were marked on the left side of the thorax with colour to monitor the age they had emerged and marked on the right side to represent the functions they were carrying out.

From these observations, Dr Allen discovered that bees from day 1 to day 21 after emerging could feed the queen but that it was mostly bees aged between day 6 and day 12 that fed the queen, the same age range as nurse bees, Dr Allen concluded that the queen was most probably fed brood food.

She observed that the queen laid eggs in batches of 15 and then rested and that this was the period when most feeding and grooming took place and that there were many more attendant bees around the queen at this time. Most of these attendant bees did not touch the queen but pointed their antennae towards her while others fed and preened the queen. She postulated that some form of communication was going on with the bees that were pointing their antennae towards the queen.

She found that feeding the queen peaked around mid-June.

In the hive she noticed certain bees were more active than others and that some bees did nothing for long periods perhaps conserving energy when their work was not needed or required , while other bees patrolled the hive carrying out jobs that were necessary.

Dr Allen observed shaker bees and carried out studies on this observation.

She observed that a bee would start shaking another bee in an up and down motion, then the bee would miss a few bees then repeat the same with another bee. She observed/ found that no shaking occurred at night or at day break. The shaking started at around 10 am and continued until the evening. She found that little shaking occurred through the winter.

She also observed that the Queen was shaken – this started in May and reached a climax before supersedure or when swarming took place.

She also observed that Virgin Queens were shaken before mating and that the shaking increased until mating occurred then gradually stopped about 1 week after mating. She postulated that shaking was a form of communication and most likely a mechanism to stimulate flight.

Part 2 of her thesis was on *Respiration Rates of Honeybees*.

The first paper I can find where Dr Allen is cited is in *The British Journal of Animal Behaviour* 1955 Vol. 3 No 2 p66 -69 *The Honeybee Queen and her attendants - Observations on Honeybees attending their Queen*. Then in 1956 she wrote a paper in *The British Journal of Animal Behaviour* on *The Behaviour of Honeybees preparing to Swarm* Vol. 4(1) p 14-22. Another paper was printed in 1956 in *The Journal of Economic Entomology* 1956 vol. 49 pages 723-726

It was titled :- *The Influence of colony size and of Nosema disease on the rate of population loss in honeybee colonies in winter*. It was co-authored with Dr Jeffree.

The first publication in the *Scottish Beekeeper* was in September 1957 page 157 again co-authored with Dr Jeffree and titled - *An Optimum Wintering Size for honeybee colonies*.

In the May 1961 edition of the *Scottish Beekeeper* along with Dr Jeffree they published their *Simplified Method of Swarm Control*.

The system was based on removing 3 frames of brood including eggs and young larvae without bees and putting them into a new brood chamber which was made up with frames of drawn comb or foundation. The original brood chamber was then set aside and the new brood chamber put on the original floor/position of the hive. A division board was put on top of this chamber, with the entrance to the back of the original hive entrance. The flying bees would then go to the lower brood chamber with the 3 frames of brood and rear a new queen. The colony in the top box would be depleted of flying bees so the chance of swarming was greatly reduced. In the autumn the two colonies were united, with the chosen preferred queen, in preparation for the winter.

In 1964 in the January edition of *The Scottish Beekeeper* she carried out : *A Review of Some Recent (Bee) Research Work*.

As well as looking into the seasonal fluctuations in the incidence of Acarine and Nosema Disease, Dr Allen was also involved in the research of Drone Behaviour and the Seasonal fluctuations of Drone Brood.

When Dr Jeffree resigned as convener of the SBA Research Committee in 1961 (He moved to England at this time), she was co-opted onto the SBA Research committee and was an elected member of this committee until 1965. She was a member of the SBA from 1962 until 1965.

After this period, she got married and became Dr M Delia Seager. On the sudden death of John Murray of Hamilton in 1965 she became the Official Microscopist of the Scottish Beekeepers' Association (Oct 1965). Delia Seager was born in1929 and died November 2010 (In the Stoke / Staffordshire Area). She got married on the 2nd of April 1965 in Aberdeen and moved to Kinlochleven.

(Mrs Delia Seager of "Cullaig" (Edencoile) ,Kinlochleven, Argyll. Was an Annual Associate member of the SBA from !966 to 70 but not in 1971 onwards, this is when she probably moved to England).

Delia Seager had a daughter Marian in 1967 and two years later her husband died.

Delia Seager from January 1970 to August 1977 was an Associate Editor of the Journal of Apicultural Research part of the Bee Research Association.

In 2013 in conjunction with Aberdeen and District Beekeepers Association - The Delia Seager Fund was established from a donation by the Aberdeen Friends Church for Delia Seager's professional interest in beekeeping. The Money was used to buy a number of the SBA booklet "An introduction to Bees and Beekeeping" which were then given to new beekeepers attending the ADBKA beginners course.

Dr Allen made a significant contribution to bee research and beekeeping and it is important that we do not forget this contribution or the work she carried out.

Margaret (Peggy) Logan

Fig 13.1 Photo of Margaret Logan

Margaret Logan (Peggy) was Assistant Lecturer in Beekeeping, North of Scotland Agricultural College. Margaret Logan was introduced to beekeeping by Dr John Anderson, in fact she was one of the girls he taught beekeeping at Craibstone. She was a great supporter of his method/ system of beekeeping and worked Glen Hives, the hive he had designed. It was John Anderson who appointed Peggy Logan as Assistant Lecturer at Craibstone. She retired in 1972 and died suddenly in 1973 while attending a beekeeping Conference in Argentina.

She was credited with the eradication of American Foul Brood in the Nairn Area following the process devised by A S C Deans where bees were dusted with pulverised Sulphathiazole.

She was awarded the Dr John Anderson Memorial Award in 1969.

Margaret Logan bequeathed £100, all her bees, film slides, films and equipment to the SBA and her library of books went to the Moir Library.

Peggy Logan was co- author of the book – *Beekeeping: Craft and Hobby*. The other author was A R Cumming.

Alex R Cumming

Fig 13.2 Photo of Alex R Cumming

Alex Cumming had moved north on retiring from being Rector at Kilmarnock Academy to concentrate on his bees and garden. He wrote / edited the Book *The Northern Beekeeper : The Hand Book of the Inverness-shire Beekeepers' Association*. Unfortunately, he died before *Beekeeping: Craft and Hobby* was published.

During his time at Kilmarnock Academy, from 1926 until his retirement in 1938, Kilmarnock had a number of well know beekeepers, Johnnie Walker in Kilmaurs, James Struthers, Ernest Ling and John Hood in Newmilns, Joseph Tinsley , James Cochran and Major Yuile in Kilmarnock, to name a few.

For those readers who know Kilmarnock, they will know the closeness of the old Kilmarnock Academy on Elmbank Drive to the Old Technical College on Elmbank Avenue where Tinsley was based and The Dick Institute, opposite the Technical College on Elmbank Avenue, where Major Yuile had a honey bee observation hive in the 1930s. I have often wondered if the three ever met up to have a cup of tea and a bee chat. Just think what association meetings must have been like with these renowned beekeepers present.

Alex R Cumming was the first beekeeper to be awarded the Dr John Anderson Memorial Award in 1945. *Beekeeping: Craft and Hobby* was written as a beekeeping guide book for hobby or small-scale beekeepers (The book was published after his death). It was written so that it would be suitable for all beekeepers in Britain and was based on Cumming and Logan's own beekeeping experiences and manipulations which had been tried and tested over many years.

The book is a wealth of information and is well worth a read if you can get your hands on a copy. For example, did you know that when Caesar came to Britain in 55BC the local inhabitants were already keeping bees and making an alcoholic drink from honey and wheat.

There are chapters including the natural history of bees, hives and equipment (P 37 fig 13 of a clearer board, I wonder if Mr Horsley had an aha moment if or when he saw this board), seasonal management, swarms, nuclei and synopsis of good beekeeping practice.

Fig 13.3 Drawing of clearing Board

Bernhard Mobus

Fig 13.4 Photo of Bernhard Mobus

Bernhard had been a German PoW during WW2 who did not return to Germany at the end of the war. While living in Lincolnshire he decided to take up beekeeping. He read beekeeping books for a year, bought a hive the following spring, made an artificial swarm in May and had two hives and 136lbs of honey by the end of the year.

Bernhard Mobus followed in the footsteps of John Anderson, Alex S C Deans and Margaret Logan working as Beekeeping Advisor at the North of Scotland College of Agriculture, running the beekeeping unit at Craibstone from 1974 until his retirement in 1990.

Bernhard wrote extensively in *The Scottish Beekeeper* in the 1980s on many subjects – Queen Rearing (1980 p 55), Feeding Bees (1981 p 9), Swarming (1982 p 76.) the Swarm Dance and other Swarm Phenomena, a must read for those who want to read a colony from the hive front (SB 1989 p 73, p94, p112, p146) and Varroa. (1983 p 101.)

He wrote letters to the *Scottish Beekeeper* about articles or comments made in the *Scottish Beekeeper* giving balance and an insight into the points being made, allowing normal beekeepers to be better informed and achieving a greater understanding of the aspect of beekeeping being written about.

He was co-author of the *Varroa Handbook* with Larry Connor and the *New Varroa Handbook* with Clive de Bruyn which had the catchy strapline of "Dead Bees Make No Honey", suggesting you had to be up to date in dealing with varroa so you could guarantee your next years' honey crop.

Bernhard also wrote the wonderful book Mating in Miniature a guide to queen rearing using mini nucs - another must read if you are going to rear queens and get them mated using mini nucs..

Fig 13.5. Cover of Mating in Miniature

Bernhard carried out research on the wintering of honey bees and identified there was a need to keep bees that were adapted to the Scottish Climate and was a great supporter of native or near native honey bees.

Bernhard Mobus found an old beekeeper from Maud who kept bees that he believed were ideal for the Scottish climate. They were dark, wintered well and were extremely gentle - characteristics that were much sought after. He propagated and distributed this Maud Strain of honey bees throughout Scotland during his time at Craibstone. On speaking about the bees many beekeepers saw an opportunity to improve their own stock by getting a Maud queen. No doubt Bernhard initially would be asked by beekeepers listening to his talk, if they could buy a nucleus of bees or a queen when the talk had finished. From these requests he set up a system of preparing and distributing queen cells, queens and nucleus colonies throughout Scotland. I have not met many beekeepers from the 1980s who did not secure a Maud queen or descendent queen from that period.

To maximise the distribution, in 1982 Bernhard set up the scheme at Craibstone to supply beekeepers with queen cells from day old larvae, grafted and cared for 24 hours. The grafted larvae from his selected queen(s) were put in a queen-less starter colony. After 24 hours when the bees had started to draw out the queen cells, the queen cells were placed in polystyrene blocks (They had been drilled to accept the queen cells). Before the cells were placed in the blocks the holes were filled with hot water which was then tipped out and the queen cells immediately placed in the holes of the block. The Cells in the polystyrene blocks were then picked up by beekeepers who took them to their own apiary where they were given to finisher colonies that had been prepared prior to the collection so that the cells could be introduced immediately on arrival. The biggest issue was making sure the queen cells did not dry out rather than from chilling from the cold.

Others collected Queen Cells from Craibstone or collected them from Bernhard after an Association talk.

I would expect most beekeepers in Scotland especially those in the North East will have bees that have some of the genetics of the Maud strain in their colonies.

Bernhard retired to the South of France in 1990 and died in 2004.

He was awarded The John Anderson Memorial Award in 1990 and awarded The National Diploma in Beekeeping in 1970, he was only one of six people who had been awarded the diploma between 1969 and 1982.

His contribution to beekeeping was great both from a practical basis and in the challenging and improving the understanding of beekeeping thoughts at that time. He helped beekeepers improve the quality of their bees.

Bernhard was sadly the last Beekeeping Advisor at the North of Scotland College of Agriculture in Aberdeen.

James Savage

Fig 14.1. Photo of James Savage carrying a crate

James Savage NDB lived in Inverness for almost 20 years, dying in 1985.

He was a Tarbolton boy who had learned his beekeeping from his father who kept bees. James was well known in beekeeping Associations throughout Scotland particularly the South West as he was employed at Auchincruive, West of Scotland Agricultural College, initially as a lecturer of Beekeeping and then Head of Beekeeping, he retired in August 1966.

Fig 14.2 Photo of James Savage receiving Retirement Gift from Ayr BKA

During his time at Auchincruive he had gained a National Diploma in Beekeeping, had written articles for the *Scottish Beekeeper* and had given numerous bee talks/lectures to Beekeeping Associations. Nothing out of the ordinary you may think.

However, James Savage was no ordinary beekeeper.

In 1922 James had joined the Cameron Highlanders, being awarded the Military Medal for bravery in Palestine in the 1930's, as well as becoming a Company Sergeant Major for the regiment.

Fig 14.3 Photo of Company Sergeant Major Savage

During WW2 he was captured and taken prisoner at St Valery en Caux on June 12th, 1940 when the 51st Highland Division had been left behind following the Dunkirk evacuation at the end of May 1940. He endured a forced march across Europe and Poland but was sent to a Punishment Camp for Non-Commissioned Officers, Stalag 383, near Hohenfels after refusing to work in Munich clearing up bomb damaged areas.

James heard of a beekeeper who lived near Stalag 383 and managed to persuade the Kommandant to let him visit the beekeeper, who was a local farmer. The farmer agreed to lend James a colony of bees on the condition that they were returned at the end of his time at the camp. The farmer gave them all the equipment they needed to get started, a hive, frames, foundation and other equipment.

Fig 14.4 Photo of Apiary at Stalag 383

The Captive Drone Society was set up at the camp and James trained many of his fellow prisoners as beekeepers.

Fig. 14.5 Photo of Members of The Captive Drone Society

Some of whom like George Lochtie from Aberdour continued beekeeping when they returned home, another PoW at Stalag 383 and member of the Captive Drones' Association was George Munro who was the Secretary of Easter Ross Beekeepers' Association for 21years.

Fig 14.6 Paper clipping published when James passed his BBKA Craftsman examination.

Apparently, James set up an observation hive in the dentist's room in the camp hospital to create a welcome distraction for the prisoners waiting to have teeth extracted.

On one occasion when one colony swarmed the whole camp turned out in force to watch. James worried that he would lose the swarm started to walk slowly through the bees, moving towards a large rock in the centre of the camp. He was hoping that the queen would follow him and try to settle near him on the rock. Fortunately, the queen and bees settled there as planned, thus adding to the myth the guards and PoWs had, that James had some unseen power over the bees.

He was able to get books through the Red Cross and YMCA and he used Red Cross crates to make hives which he later used to collect and house swarms that were collected. There is a photo in the September 1944 *Scottish Beekeeper* with James Savage and Captain Reverend Kenneth Grant from Fort William (Later to become a Bishop).

Fig 14.7 Photo of James and Kenneth Grant.

James had just passed the Craftsman Examination of the BBKA (The Beemasters Certificate was the SBA equivalent).

The BBKA supported the prisoners every way they could and this including sending out exams via the Red Cross.

The PoWs only got a little honey as most was left for the bees along with additional sugar taken from the PoWs rations to allow the bees to survive the harsh winters.

The whole camp was fascinated by the bees which flourished and by the end of the war had grown to 4 colonies which as promised were returned to the farmer.

Keeping bees had a positive effect on the camp, keeping the PoWs busy during their time there, as well as being good for their mental health.

The Kommandant was happy to support the PoW beekeepers and beekeeping within the camp as it kept the PoWs from getting bored and therefore less likely to try to escape.

In 1945 when the camp was being moved from Hohenfels to Landshut to avoid the advancing American and Russian Armies, James managed to escape, hiding in woods for four days before meeting up with the advancing Americans.

When he left the army after the war he went to Auchincruive and followed a career in beekeeping, a job he dearly loved.

He retired on the 31st of August 1986, moving to Inverness.

A Brief History of Scottish Beekeeping and Beekeepers

Fig 14.8 Presentation by LDBKA on James Savage retirement

James Savage was not the only notable Scottish beekeeper during wartime.

Mr John Fyfe of Springhill Brae, Crossgates, (A well known local beekeeper) at a meeting of Dunfermline and West Fife Beekeepers gave a talk in March 1961. He spoke about his beekeeping exploits during WWI.

John Fyfe helped French farmers who kept bees and on occasions sent packages of bees from Flanders home to Scotland. Apparently, the bees passed through without being opened by the censors.

John Fyfe spoke about collecting a swarm that was hanging on the barbed wire at the front line. The swarm had attracted a lot of attention from troops in the area and John had to warn them to keep their distance. Although he made the warning on a number of occasions it was only when the bees started to fly the troops made a hasty retreat. John Fyfe describing it as the fastest retreat made by the British Army in all its history.

I am sure there were other beekeepers who had notable war experiences such as Dr David Christison.

Fig 14.9 Photo Dr D Christison

Dr Christison was awarded the SBA's Dr Anderson's Memorial Award in 2000 and was one of the camp doctors when a PoW on the River Kwai or Dr Suzanne Ullman a lecturer in Zoology at the University of Glasgow, who was born in Budapest, whose parents were Jewish. Suzanne had a harrowing time as a child evading the Nazis.

We should never forget their stories and the difference that they made to other peoples' lives.

Captain Leonard MacDonald Thake

Fig 15.1 Photo of Captain Thake

Captain Thake died 5th April 1977 aged 81 (Born in 1895). He started beekeeping in 1909 keeping between 150 and 200 colonies for honey, breeding and experimenting. He favoured the native black bee as well as the French Le Gatinais bee. He wrote prolifically from the first Edition in 1924 of the *Scottish Beekeeper* until 1976, just before he died.

He was made an Honorary Vice President of the Scottish Beekeepers Association in recognition of his work and he received the Dr John Anderson Memorial award in 1964.

He was a lecturer in Beekeeping at the East of Scotland Agricultural College in Edinburgh, was a past President of the Fife Beekeepers' Association and had been president of BIBBA.

As well as writing profusely in *The Scottish Beekeeper* for over 50 years he was an SBA examiner and honey judge. His certificate number for Honey judge was number 28 (John Anderson was No1).   Not only was he well known, he knew all the great beekeepers of that time. He had worked for Mr FWL Sladen, at Ripple Court Apiary near Dover, as what he described as his "gofur". During World War 1 he served as a lieutenant in the Highland Light Infantry. After the war he became Chief Apiarist, Dominion of Canada, Department of Agriculture, living at Bran End, Stebbing, Essex before being appointed assistant instructor in beekeeping at East of Scotland,

Agricultural College and by 1924 was a commercial beekeeper breeding and selling bees from Dura Den, Cupar, Fife. During World War 2 he was promoted to Captain in His Majesty's forces.

His articles covered most areas of beekeeping with a philosophy of keeping - beekeeping simple and fun. He not only wrote under his own name but that of El Emtee the Scribe and although humorous it contained some pearls of wisdom.

In one it advised beekeepers to beware of false prophets and the following of their teachings. This is sound advice even today. I am not going to say I am immune from these beekeeping seers; however, I have heard of a beekeeper that my friend and bee buddy Sandy Cran calls a butterfly beekeeper (because they continually flit from one idea to another) who after reading or hearing a bee talk on feeding bees decided to stop giving their bees a winter feed of ambrosia/syrup. The easy way was to put a 7.5kg block of fondant on the top frame bars of the brood nest. Unfortunately, after the winter the beekeeper had few hives with living bees in them. The beekeeper's response was one of goodbye to a load of rubbish and that a cull was necessary to keep their colonies strong. Perhaps their bees were a lot of rubbish and the advice good, but I am not so sure. In most cases it is easy to blame the bees and make it their fault and that it has nothing to do with them - the beekeeper. Is it possible that the beekeeper had not appreciated some aspect of the process/system given in the advice on winter feeding and had therefore failed to carry out the manipulation/requirement necessary to look after the bees. Perhaps there was a requirement where frames of honey should be left for the bee colonies. I know of some beekeepers who take every ounce of honey out of their hives in August/ September leaving nothing but syrup to compensate the bees for the honey they have taken. (Syrup is significantly cheaper than honey so why would you leave the honey when an easy substitution can be made? That is their thinking)

Recently I have seen two experienced and highly respected beekeepers (by me) offer words of caution on certain beekeeping practices which were being made via a group chat/ blog / online, only to be shot down and their advice disregarded.

As the saying goes - *if you believe everything you read you will eat everything you see*.

Perhaps when we read these things online or hear of some new wonderful manipulation, no doubt given in good faith I am sure, there should be a **Bee Warning - this may damage your bees' health**.

Captain Thake's advice to beginners was to buy and follow the guidance of a good beekeeping guide book and to focus on the factors that govern a bee colony throughout a season and to keep to methods that are close to the honey bees' natural cycle as possible.

One subject Captain Thake did not cover in his writings was Telling the Bees, although he did cover most aspects of beekeeping. However, in 1969 he wrote a letter in *The Scottish Beekeeper* asking for information from readers about Telling the Bees and Tanging. The letters the *Scottish Beekeeper* received were very interesting but I am not aware Captain Thake ever published an article on Telling the Bees. He did write about Tanging a Swarm in Random Notes p 153 of the *Scottish Beekeeper* in 1965, when he wrote about his friend James Kirkcaldy tanging down and collecting a swarm of his bees.

The folklore goes back a long way in time, perhaps to Greek / Roman times (The Greeks Aristotle and Virgil wrote about tanging) or maybe even further back perhaps back to when the Greek God Aristaeus was teaching humans how to keep bees.

Tanging

It was thought /an old custom that bees swarming settled quicker if the bees could hear a loud banging or ringing noise and thus make it easier to collect the swarm. Butler wrote in his book *Feminine Monarchie* there were two reasons for tanging, the first was to let others know that there was a swarm of bees in the air and that the person who owned them was following the bees, the other reason was to stop / interfere with the communication between the queen (through piping) and the flying bees so that the swarm would settle rather than fly a great distance before settling.

Although there is no evidence that tanging works, Alan Riach a Past President of the Scottish Beekeepers Association tells a story about one of his late brother's hives swarming. Just as he watched the swarm take off and the bees were making their getaway, two buccaneer jets from the local RAF station (Lossiemouth I think) carrying out low flying exercises passed over. As the two planes broke the speed of sound, two extremely loud bangs could be heard, the bees instantly stopped in their tracks and returned to their original hive. So maybe there is some truth in the matter, it's just an extremely really loud noise / bang that is required and as Allan said not everyone can arrange the suitable tanging with buccaneer jets at swarming time.

People perhaps, as Butler said, tanged their bees so they could not be accused of trespass as they followed the escaping swarm which belonged to them and really had nothing to do with the swarming bees settling. Apparently as you tanged you could not be accused of trespass as well as being a good way to tell people that the swarm came from your hives.

Telling the Bees

This old custom involves telling the bees of important events which impact on a household like births, weddings and funerals, if this was not done it could lead to bad luck and the possibility of the bees not producing honey, flying away, not thriving very well and even colonies dying out.

In 2022 on the death of Queen Elizabeth 2nd her bees at Buckingham Palace and Clarence House were told of her death by John Chapple the Royal Beekeeper. Black ribbons were tied in bows around the hives and the bees were informed of the Queen's death and that King Charles would be their new master. John Chapple urged the bees not to go, to be good to their new master and that King Charles would look after them well. (He already kept bees at Highgrove and Dumfries House)

There can be variations to the custom, for example in some places the hives would be tapped (sometimes with the house keys) or lifted (sometimes the hives being turned round a full 360 degrees) at the same time as the coffin was being lifted into the hearse and the bees being told of their master's death. In other places the bees were told quietly in the middle of the night.

For a death the hives would be covered with ribbon or a black cloth and funeral cake and wine placed on or near the hive.

The bees should be told of births within the family/household and should a bee settle on the baby it was thought that this would bring the baby good luck. Saint Ambrose the patron saint of beekeeping, around 340 AD had honey comb built near the cradle he lay in by honey bees and this was seen as a sign of good omen.

With a wedding in the house, the bride to be should visit the hives and "tell them her story" (the details of who was getting married) before going to the church or where she was going to be married. The hive should be approached from behind by the member of the household getting married and they should bend over the hive, tap it gently and say " Dear bees I am going to be married." If this was not done they ran the risk of losing the bees.

The hives again should be covered with a colourful cloth/sheet or colourful ribbons and wedding cake left.

If hives were being kept by newlyweds at their new home, they had to visit the hives and introduce themselves to the bees as it was believed their married life would be unlucky if they did not.

I have read somewhere that telling the bees was about ensuring the well being of the bees. If someone takes the time and trouble to tell the bees what is happening then those same people will take the time and trouble to look after them and the bees will prosper. The reverse being true - if someone does not take the time and trouble to tell the bees. They will not look after them and the bees will do badly, possibly swarming or even die out.

Although Captain Thake did not write an article on telling the bees he wrote on most aspects on Practical Beekeeping over more than 50 years, giving guidance on how to improve beekeeping practice and improve your stock of bees . He freely passed on his experience and knowledge and promoted having fun and keeping things simple when beekeeping.

Mr. L. M. Thake showing the Drumming of Bees at the   combined wit

Fig 15.2 Photo Captain Thake Demonstrating Driving Bees

Captain Thake's words still resonate as much today as they did during his lifetime, he was a Great Practical Beekeeper and we should remember him and what he said.

The Shepherds

Mr and Mrs Shepherd of Greenfield, Newton Mearns.

The Shepherds did not come from a family of beekeepers. Mrs Shepherd was offered a hive and stock of bees from a neighbour who was moving home. She bought the stock and hive for £1 and that was the start of a lifelong hobby for both she and her husband Willie.

The initial hive was a double walled hive however latterly they were to work single walled hives which they found easier to manage.

The shepherds were well read on beekeeping with Mr Shepherd designing a number of appliances and hives during his time as a beekeeper. Mrs Shepherd said that Mr Shepherd had his own ideas on managing colonies and his methods were based on what he read, what he thought were the key actions, adding bits and taking bits away to form his own system / method of beekeeping.

Over the years he produced his own method of Swarm Prevention and his own method of running multi queen hives to generate extremely strong honey producing colonies.

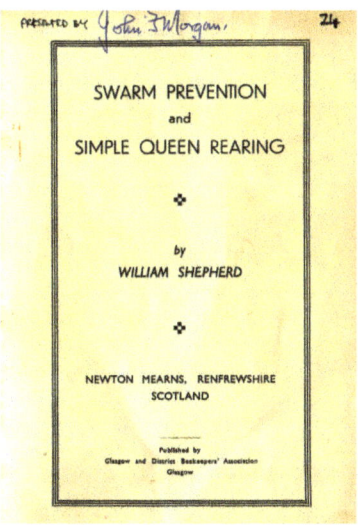

Fig 16.1 Front cover of pamphlet on Swarm Prevention

His method of swarm prevention /control was :-

The Queen, empty drawn comb, comb with stores and frames of foundation were placed in a brood chamber at the bottom of the hive and therefore along with the flying bees. Above this was placed the queen excluder, any supers with honey then a shepherd board with Shepherd tube and a brood chamber with brood frames and any queen cells.

Fig. 16.2 National Hive with division board and Shepherd tube

The bees in the top being fed with syrup. Young bees were allowed to move up to the top brood chamber over a period of 3 days then the shepherd board was closed off so bees in the top chamber could only leave via the shepherd tube which exited only a few centimetres from the bottom entrance. On returning to the hive those bees exiting via the tube entered the hive via the bottom entrance and thus depleted the top box of bees reducing the chance of the top box swarming. When a new queen emerged from a queen cell it would exit the chamber via the tube and on its return enter the bottom brood chamber and supersede the old queen.

His method of running multi queen hives was based on the method published in the book *The Sky Scraper Hive* written by Father Dugat and published in English in 1948 but with modifications.

Dugat's system involved setting up a number of colonies one on top of each other, then giving the bees additional space - brood boxes and frames, sufficient for brood expansion and additional food stores to inhibit early swarming. Two weeks before a major honey flow the queens were removed and the frames of brood consolidated and separated into brood boxes of unsealed and sealed brood. The unsealed brood put at the top to make it easier to find queen cells and remove them. By managing the production of queen cells and keeping the bees queen-less for approximately

30 days (but no more) helped produce a very strong colony of mainly foraging bees, available to take advantage of the honey flow. The bees would collect nectar and produce high yields of honey, filling the brood boxes with nectar and honey as the brood emerged. The brood boxes would be full of honey in just a few weeks. At the end of the honey flow the boxes were moved a distance away from the original site and a queen and small nucleus in a brood chamber along with additional boxes above this chamber put on the original site. Most of the bees - which were foragers returned to this hive. Any bees left on the frames of honey were brushed off and the honey then harvested. During the period of queen-less-ness, queen cells could be produced and harvested to rear queens.

The Shepherds worked hives with 5 colonies/queens, each queen in a brood box stacked on top of one other. They found the hive with 5 brood boxes was too high when additional space/brood boxes were added. So, at this stage, they went to three colonies plus additional brood boxes on the original site, with the other two queen brood boxes along with any added brood boxes placed behind the original hive. The colonies were allowed to build up. When the colonies had built up, the brood boxes with queens were stacked in their brood boxes facing in the opposite direction and at the back of the original hive which was set up with a single queen in a brood box along with all the brood boxes that had no queens in them above this box. When the brood boxes at the back were again full of bees the brood boxes and queens were moved so that the flying bees would fly to the original hive through entrances on the back of the brood boxes on the hive on the original site. This generating an extremely strong colony which was capable of producing a lot of honey when the conditions were right, just before an expected honey flow.

Mr Shepherd designed hives – winning two seconds prizes at the Highland Show for hives he designed. He also designed the gimp pin feeder which was sold in America by Gleanings.

Mr Shepherd designed the Shepherd tube and the Shepherd board which was part of his method of swarm prevention/ control.

The shepherds were mentors to many beekeepers and their apiary was open to all beekeepers to visit. Mr Shepherd was the beekeeper of choice to do hive manipulations at apiary visits.

Mrs Shepherd thought "Weather was everything in Beekeeping" and that you should not manipulate bees in poor weather. The Shepherd's hives were sited so as not to face the wind.

Mrs Shepherd thought that when working bees, the work should be planned and the beekeeper should be able to cope with what they found in the hive. She felt beekeepers should have the ability to change plans accordingly to what was happening in the hive.

She believed that experienced beekeepers could read a colony from outside the hive and that when the colony was expanding the bees should be left alone but if the bees were dwindling or lazy and loafing around, an inspection was required.

That on a regular hive inspection if there were two or three empty rows of cells above the brood the bees needed to be fed and if stores were pressing onto the brood that more space was required for the bees.

Her advice to new beekeepers was to learn to be patient and handle gently and firmly, taking time when handling bees.

Her plan for September was to ensure the bees / colony had sufficient stores to survive the winter along with a couple of frames with or without brood to allow for brood laying in the hive.

She believed beekeepers should gain knowledge by watching the bees entering and leaving the hive – at what speed did the bees enter and leave, were they loaded with pollen, were they carrying nectar i.e. legs hanging/spreading to the back. If bee legs were hanging forward they were not carrying nectar.

(Remember young bees do orientation flights, generally around midday – and you will be able to tell if the colony is expanding or contracting from observing the number of bees doing orientation flights as well as the number of bees entering and leaving the hive. If you want to learn more about reading a hive, read the section by Ian Craig in the book *The Glasgow Beekeepers*. Remember you can also tell what is happening in a hive from the sound the bees are making as well as from what you smell.)

In 1936 Mrs Shepherd was secretary of Glasgow and District Beekeepers' Association (GDBKA) and Mr Shepherd was treasurer.

In 1938 Mrs Shepherd helped support GDBKA at the Empire Exhibition Show, where she manned the observation hives and it was at the show she and Willie got the idea of starting the Glasgow Beekeepers' Club. She had worked at the show for 6 months enjoying the social aspect of meeting and enjoying the company of beekeepers from all over the world. The beekeepers met up night after night during the show. After the Exhibition show had ended Mrs Shepherd set up a Beekeeping study group which was very successful and made her think again about developing a beekeepers' social club. It was a club where you could meet other beekeepers, to have a chat, exchange ideas and buy and sell bee appliances. Unfortunately, because of the start of WW2 places to meet were hard to find. It was not until 1942 when through Mr Chalmers on Buchanan Street that they got the room to set up the Beekeepers' Club. The Club opened on the 22nd of May at 213 Buchanan Street in 1942.

The Club had moved to 104, Renfield Street by 1946.

IN 1953 Mrs Shepherd turned 70 and as an appreciation of the work she had carried out, the club presented her a box of 70 coronation shillings .

Fig 16.3 Photo of 70 shillings presented to Mrs Shepherd

Mr Shepherd was made Honorary President of the Glasgow Beekeepers' Club in 1958.

William Shepherd died in April 1967.

The Club was very lively but depended on volunteers to make it work so in 1975 a year after the death of Mrs Shepherd the club merged with Glasgow and District Beekeepers' Association.

The club had successfully promoted beekeeping in the Glasgow Area. It was through the club that the Kelvingrove Museum and Art Gallery got an observation hive. It was installed in May 1959, and the observation hive has Introduced honey bees to young and old since then. Apparently, it is one of the most popular sites within the gallery and Charles Irwin should be thanked profusely for the time he has spent ensuring the presence and care of the bees at the museum over the years.

Mrs Shepherd died in 1974 and in her obituary October 1974 *Scottish Beekeeper* it states she was a recipient of the Dr John Anderson Memorial Award in 1945 (The same year as Alex R Cumming) as well as being made an Hon. Vice President of The Scottish Beekeepers Association in 1973.

The Shepherds loved their bees and were great ambassadors for beekeeping in Scotland.

W.W. Smith and the Smith Hive

Fig 17.1 Photo of Willie Smith

After World War 1 (WW1) Willie Smith of Innerleithen was employed as a chauffeur by Mr Ballantine a local mill owner. During his spare time and with the encouragement of his employer Willie started beekeeping as a hobby in 1920. Willie had another hobby, fishing and he told James Cunningham in 1923 he hoped to make a success of one of his hobbies. He was eager to learn and read bee literature, bee books and used the Moir library extensively. He was awarded his Beemaster certificate in 1923 and became an expert Beemaster in 1924. By 1924 he had decided to concentrate his energies on beekeeping, since beekeeping had greater possibilities than angling.

He was Secretary of Peeblesshire Beekeepers' Association for many years and was a keen exhibitor at shows, winning many prizes during that time.

Fig 17.2. Photo of Willie at Honey Show

He was awarded the Dr John Anderson Memorial Award in 1946.

Around 1926 Willie Smith decided to make a single walled hive without a porch and with a flat roof. Willie decided to produce a chamber which would take 11 frames and thus be able to take the original 21 section rack (10 frame chamber being too small and a 12 frame chamber being too large). He made his hive chambers by cutting lengths of 8 $^7/_8$ inch by $^7/_8$ inch Baltic white pine which he nailed together. 4 pieces of wood rather than the 8 pieces of wood which make up a modified National Hive brood chamber. Like the American Langstroth hive he made it with a top bee space and the frames similar to British Standard deep frames but with shorter lugs similar to Langstroth frames. Smith originally called his hive the Traquair Hive as that was the district where he did his beekeeping.

Fig 17.3 Apiary at Traquair

It was later to be called the Smith Hive by Neil Anderson who drew up the Blue print drawings for the hive in 1942 and that is what it has been known as since.

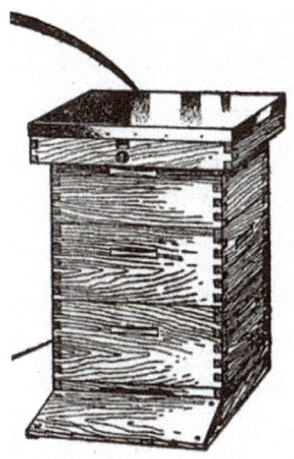 

Fig 17.4 Drawing and photo of Smith Hive

Willie Smith's hobby had become a successful side line and in 1934 he became a fulltime beekeeper initially with 50 hives, expanding to 150 by 1945 and making a living solely from honey production.

Willie Smith along with his son designed the Combined Cutter and Scrapper for removing heather honey from the comb.

Willie Smith wrote part of the Focus on articles published in the *Scottish Beekeeper* magazine in the 1950s. These articles covered Early Spring Management, Apiary Equipment, Larger Brood Chambers and the use of double brood chambers to achieve higher yields of honey and better swarm control, ventilation, queen introduction and moving bees to the heather.

He advocated that bees should be fed honey and should only be fed sugar syrup when the natural food was not available. He suggested that hives should be protected from winds to maintain colony heat and better overwintering. He wrote about the best comb should be in the top box, with old comb, drone comb and comb full of holes being gradually worked to the outside of the bottom box and then removed. He avoided interfering with the arrangements of brood in the brood chambers allowing the bees to build up naturally - he did not reverse his brood chambers and his advice was not to inspect /examine stock too early in the season because of the risk of the bees balling the queen.

He suggested that feeding bees candy excited them unduly - he also observed that bees never seal a hole anywhere in the hive other than at the entrance if the hole was large enough to allow a bee to pass through.

His management system included clipping the queen and doing a weekly or nine day colony inspection when drones were flying freely. Smith had adopted a double brood chamber system based on that of George Demuth.

In 1968 the year before he died he wrote in the *Scottish Beekeeper* his article on "How I work my Bees".

He worked a double brood chamber system, using the same size of chamber for collecting honey surplus to allow for greater flexibility. It allowed for the frames and the chambers to be interchangeable. He must have been a strong man since it is not easy to lift a chamber with 11 British Standard frames full of honey.

Bees were wintered in double brood chambers with plenty of stored honey, pollen and a good queen. By doing this he had good strong colonies in the spring.

In the spring when nectar was being gathered by the bees he added a third chamber preferably with drawn comb and once the queen was seen to be laying eggs and brood was being produced, the queen was shaken/ placed on one the 2 bottom chambers and a queen excluder placed between the second and third chambers. All drone brood being destroyed to stop the drones becoming trapped by the excluder when they emerged.

Keeping brood above the excluder attracted and kept bees in this chamber, reducing congestion below and the risk of swarming. Honey Supers could be added as required and always using a comb or combs of honey from the chamber below to attract bees into the new super.

9 days after the queen had been moved down a check for queen cells was carried out on all frames with brood, including those above the queen excluder (sometimes queen cells can be produced on combs separated by a Queen excluder). A 9 day check for queen cells was then repeated routinely along with a check that the brood chambers were not getting clogged up with honey (This would create a barrier and restrict brood rearing due to a reduction of space for egg laying and thus create conditions for swarming).

Removed combs of honey were replaced with empty drawn comb when possible. Also combs high in drone brood/comb were replaced with good comb with worker cells.

When doing a hive check / inspection, a puff of smoke was applied at the entrance of the hive and the joint between the two brood chambers broken. The upper chamber was eased forward by about 4 inches and then the top box was tilted up off the bottom brood chamber. Smoke was then blown up and down both chambers, this allowed the bottom of the frames on the top brood chamber and the top of the frames on the bottom brood chamber to be checked for queen cells in both chambers.

Fig 17.5. Checking for Queen Cells

If queen cells were found early in the season and nectar was being collected the colony would be split (see process below for making increase). If the queen cells were found late in the season a split would impact on the honey harvest and so in this case all queen cells would be destroyed and action taken to avoid swarming and remove the swarming impulse. The swarming impulse was usually caused by congestion in the upper brood chamber and by removing combs full of honey with empty comb the urge was removed. The combs of honey removed could be placed in a chamber above the queen excluder or stored and used in colonies that did not have sufficient stores for the winter.

Queen Rearing and making colony increase.

When making increase, Smith used colonies strong in bees i.e. bees crowded in 3 brood chambers with sufficient combs with honey to meet the needs of the bees in the splits. Please note, Colony increase could be made without having queen cells.

1. All supers were removed.
2. The Top /2nd brood chamber was set aside.
3. All queen cells in the bottom brood chamber were destroyed.
4. The queen was found and put in the bottom brood chamber
5. Queen cells were only in the 2nd brood chamber

6. A spare brood chamber with frames of empty drawn comb replaced the 2nd (top) brood chamber.
7. The 2nd brood chamber covered and set to the side.
8. A queen excluder, honey supers and crown board were put on top of the bottom brood chamber.
9. The bees and queen cells went above this with the crown board feed holes covered with queen excluder to allow bees to pass up or down. The crown board/ top brood chamber also had an entrance allowing bees to fly from the top brood chamber.
10. The hive was covered with another crown board and a roof.

Smith suggested it was best to carry out this manipulation when the Queen Cells contained larvae which were approximately 2 days old. 9 days after the division/split of bees, the top brood chamber along with the queen cells was moved onto a new floorboard at a new site. Any Queen Cells that were in the original hive (i.e. the bottom brood chamber) found and destroyed.

Smith suggested that the queen cells in the top box be reduced to 1 unless the colony was really strong where it could be split into more nuclei if this was wanted.

If he did not want to weaken the colony, the split would be left on top of the original colony however the bottom inner cover had to be replaced with a floor or the holes of the crown board covered with glass or plastic sheet. If this was not done there was a risk of bees from below swarming out when the queen went on her mating flight.

When inspecting/ checking a hive and the colony was found to have sealed queen cells and the colony was on the point of swarming he dealt with this situation as follows.

1. The top/ upper brood chamber with the queen was set aside.
2. All the queen cells in the bottom brood chamber destroyed except those with eggs and young larvae.
3. Queen excluder and honey supers put above the bottom brood chamber.
4. Crown board (with entrance slot) with holes completely covered put on.
5. Top/upper brood chamber with queen and queen cells put above (with entrance on cover board open allowing bees in top brood chamber to fly)

All flying bees in top brood chamber would fly to the bottom brood chamber and queen cells in the top brood chamber torn down by the bees that were left. There was a risk of swarming if the bees were not flying freely for example due to bad weather and in this case the queen cells should be destroyed.

9 days later the queen cells in the bottom brood chamber should be inspected and the best one selected and the rest destroyed.

The old queen in the top/ upper brood chamber kept until the new queen in the bottom brood chamber laying. If all was satisfactory the old queen could be killed and the 2 colonies united using the paper method. If there was a problem with the new queen she would be killed and the 2 colonies united by the paper method.

Also, if increase was wanted the old queen and the top/ upper chamber could be moved to a new site within or out with the apiary.

According to Willie Smith this system gives the beekeeper almost complete control over his bees.

Willie Smith died in 1969 however his legacy still lives on. George Hood of Ormiston was tutored by Willie and he took over the management of Willie's bees on his retirement and then were incorporated into George's stock. The honey from these bees can still be bought from Hood's Honey.

Willie Smith's double brood chamber method of beekeeping has influenced many beekeepers since and many including myself work a similar system of managing their bees today. Willie Smith was a great practical beekeeper and I am sure you will agree was one of Scotland's Greatest Beekeepers.

William Hamilton 1889 – 1977

William Hamilton was born in Garelochhead in 1889, his family had the local joinery business which had been started by his grandfather who had moved from Ayrshire to Garelochhead in the 1830s.

William wrote two books – *Recollections of Garelochhead 100 years ago*.

Fig 18.1 front cover of Recollections of Garelochhead

This was an account of Garelochhead in the late 1800s and early 1900s. William became a beekeeper after finding hives on a bench in the family joiners' workshop that had belonged to and been made by his deceased uncle Robert. He bought his first bees for 10 shillings at the age of 14 when the manager of the local grocer's shop a local beekeeper was emigrating and with the help and encouragement from his family particularly from his aunt Maggie, William successfully kept bees. William served initially in the army and then the Royal Flying Corp during the First World War. After the war he became lecturer at the West of Scotland Agricultural College - with a focus on Beekeeping. He moved to become Lecturer in Beekeeping at the University of Leeds and then to The Yorkshire Institute of Agriculture as lecturer in Beekeeping, a position he retired from in 1955.

He was an acknowledged authority in beekeeping and wrote the practical beekeepers' book of choice in the mid to late 1900s - *The Art of Beekeeping*, which was first published in 1945 and ran to 3 editions.

Fig. 18.1.1 Cover of The Art of Beekeeping

After retiring in 1955 he lived in Canada before returning to Scotland and Garelochhead initially and then to Blairmore, Dunoon where he died. He wrote of his recollections while he was in his 80s but it was only published much later by Northern Bee Books, long after his death.

William Hamilton died at the age of 88 in 1977.

The Art of Beekeeping was first published in February 1945. It was William Hamilton's aim to produce a book that would help beekeepers become more competent beekeepers, explaining / describing the what and why of manipulations in plain / simple language.

Hamilton wrote about the Natural History of the Honey Bee. He explained how bees rest in winter , however when the sun returns Northward that egg laying begins. The kind of weather and the presence of spring flowers also influencing the queen in laying eggs. Hamilton stated that successful beekeeping was mostly dependent on the knowledge of bee behaviour and the actions and reactions of bees to stimuli both natural and artificial (i.e. through manipulations by the beekeeper). He believed that failure was generally due to the inability of the beekeeper to identify the impact of one or more factors (absent or present) on the colony and its ability to prosper.

In the book he discusses a number of types of hive that were available and the merits of the hives, frames, size and construction. He also wrote about purchasing bees and how to manipulate and handle them. He believed that by handling bees properly, created an environment that would cause little or no discomfort to the bees or the beekeeper. However, if you were rough with bees then they would be unpleasant and aggressive to the beekeeper and others.

The bees' instinct is to defend its nest but also has the instinct to preserve their own life and so the bees respect slow leisurely movements. He also wrote about bees becoming demoralized by fear, that smoke caused fear in bees, demoralizing them which then encouraged the bees to eat honey and therefore not be inclined to sting. With hives short in stores/ honey they were more difficult to handle and in such a situation, lightly feeding them would make it easier to handle the colony. He also stated that beekeepers should be in no great hurry to open the hive, to allow time for the smoke to permeate through the hive before opening, Hamilton highlighted the importance of Autumn and Winter management. He suggested that the hive should have at least 20 frames (for a National Hive) 10 x10 so that there was sufficient comb for breeding as well as for sufficient stores to get through the winter. Hamilton suggested that beekeepers should reserve a portion of surplus honey for the bees to meet their needs during the Winter and Spring. "Honey makes Honey and it cannot be too strongly emphasised that combs of sealed honey are vastly superior to sugar syrup."

The longer I keep bees the more I am inclined to agree with him. Hamilton suggests that the equivalent of 10 combs of sealed honey be left for the bees.

In his section on Spring Management, his advice /aim was to create strong colonies that could take advantage of honey flows. He also recognised that the weather in March and April influenced the progress/ development of the colony. Cold and wet weather retarded development and sunny, mild weather assisted the development of the colony.

The book includes sections on Swarm control (different methods of swarm prevention and control are mentioned), Honey production including heather honey are discussed.

There are sections on Queen rearing. Queen introduction, preparing honey for market and for show as well as a section on Pests and Diseases.

Although there is nothing about Varroa, (remember this book was written approximately 50 years before Varroa reached Britain) there is still much useful information and good advice given in this book and Hamilton's aim of a book that makes beekeepers more competent still applies.

James and George Braithwaite
- The Braithwaite Brothers

James Allan Braithwaite was born in 1907 and died at the age of 83 in February 1990. He was a member of the Braithwaite family of Dundee who own and run J A Braithwaite Ltd. Tea and Coffee Merchants and Specialists, who import a wide range of Teas and Coffee beans and who sell their goods and products from the Oldest Trading Shop in Dundee which was established in 1868 and who have traded from 6 Castle Street, Dundee since 1932.

During the mid 1900s, a number of influential beekeepers lived in Dundee and the surrounding area.

These beekeepers included Harry R Brown who died December 1963 around (I believe) the age of 77.

Fig 19.1 Photo of Harry Brown

Harry Brown started his working life as a young reporter for D C Thomson Publications. He was born in Corstorphine in Edinburgh. His career with the publisher was from 1902 until 1952. From 1915 to 1916 he worked on the Weekly Welcome magazine but later he held a number of responsible positions including editor of one of DC Thomson's publications.

Harry enjoyed beekeeping and he became actively involved both in The East of Scotland Beekeepers' Association and the Scottish Beekeepers' Association.

He wrote in the *Scottish Beekeeper* magazine under the name of "Angus" where he answered beekeeping questions sent in to him by beekeepers. Harry had suggested the idea, to Dr John Anderson who was editor of *The Scottish Beekeeper*, a magazine feature called Question Time. Dr Anderson's response was "Excellent idea, will you take it on?" The answer was yes and the column was started in 1929.

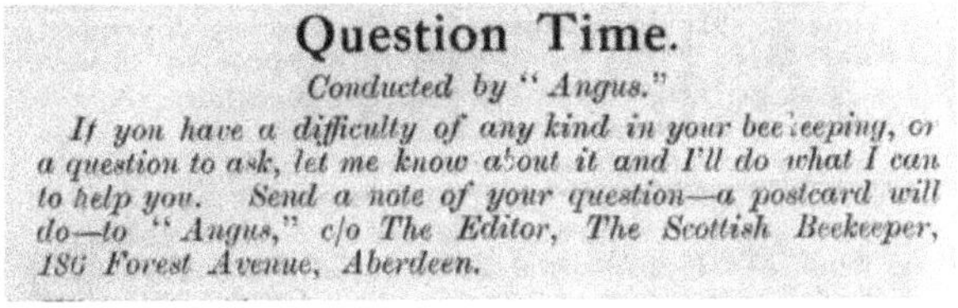

Question Time.
Conducted by "Angus."

If you have a difficulty of any kind in your beekeeping, or a question to ask, let me know about it and I'll do what I can to help you. Send a note of your question—a postcard will do—to "Angus," c/o The Editor, The Scottish Beekeeper, 186 Forest Avenue, Aberdeen.

Fig 19.2 Header for Question Time Column

He was president of The Scottish Beekeepers' Association from 1941 to 1943.

He was secretary of the Scottish Beekeepers' Association from 1960 until his death in 1963.

After his death his wife presented the SBA with £130 - a substantial amount in 1964. The income from this money was to be used to stimulate interest in *The Scottish Beekeeper* magazine and to provide awards/prizes for competitions held by the *Scottish Beekeeper* magazine.

The award was initially a photographic competition but more recently has been awarded for the best article or series of articles published in the *Scottish Beekeeper* magazine.

Others beekeepers from the Dundee area included Adam McClure who wrote prolifically in the *Scottish Beekeeper* both anonymously and under the initials of AFFM. Adam was presented the SBA Harry Brown Award twice in 1970 and 1975 and the SBA Dr John Anderson Memorial Award in 1981. He was also the SBA Library Convener from 1967 to 1980 and General Secretary of the SBA from 1971 to 1973.

Captain Thake of Dura Den, Cupar, who also wrote prolifically in the *Scottish Beekeeper* and was awarded the SBA Dr John Anderson Memorial Award in 1964.

David Robb who was best known for the observation hives he designed and made and who was presented the SBA Dr John Anderson Memorial Award in 1973.

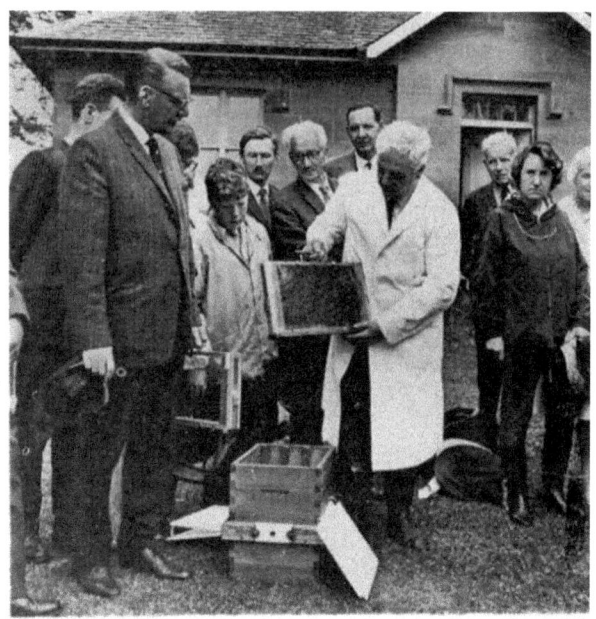

Mr David Robb, Dundee, showing his hand-carrying Observation Hive. This hive is suitable for lectures in schools, etc.

Fig 19.3 Photo of David Robb showing observation hive

Rev. David Grimmond President of the Scottish Beekeepers Association from 1963 until 1966.

Willie Wilson who was the President of Scottish Beekeepers Association from 1967 until 1969.

John Gilmour who was the President of the East of Scotland. Beekeeper' Association.

As well as his brother George Braithwaite.

James who was known by some of his friends and family as Jimmie started beekeeping at an early age, by helping his beekeeping father who kept bees at Longforgan (South of Dundee on the A90)

By all accounts James was a good beekeeper - he helped other beekeepers whenever he could – giving his time and advice freely.

13 year old Sandy Brown, Honeygreen Road, Fintry, Dundee, helping to look for the Queen. Mr James Braithwaite is on right of photograph.

Fig 19.4 James Braithwaite at Apiary Visit

He did demonstrations at his own apiary at Bridge Street, Barnhill, Dundee, as well being the beekeeper of choice in the Dundee area when demonstration visits were organised to open and go through hives. He was able to demonstrate the manipulations required to manage a colony including marking and clipping the queen. The use of clearing boards and much more, both clearly and confidently.

He gave talks/lectures at Association meetings throughout Scotland covering most areas of beekeeping. He had a particular interest in treating bee diseases, a subject he would bring up even if talking on another bee subject. He successfully treated a hive in 1947 with AFB using Sulphathiazole, writing about this in the *Scottish Beekeeper*. He also wrote about using Ammonium Nitrate in the smoker- On heating the Ammonium Nitrate produces the mild anaesthetic Nitrous Oxide – he used this method to introduce queens and unite colonies.

He gave a talk to Dunblane and Stirling "on single Brood Chamber for honey production" in which he spoke of his beekeeping system. It was based on a Scottish Black Queen which he believed laid on average at peak times 1,300 eggs per day which meant that the hive would comprise of six brood frames, two pollen frames and three for storage of honey. He felt that a single brood chamber was more suitable for migrational beekeeping. He went on to talk about the importance of checking for disease and of his methods for dealing with Acarine, Nosema and Braula. He also explained his method of cutting comb into half pound pieces of comb honey with minimum of waste.

This was the subject of his talk to Edinburgh and Midlothian in 1968 - the title of his talk was "Cut Comb and A Lifetime on Beekeeping". A similar talk was given to the Glasgow Beekeepers' Club in November 1969.

Other areas of beekeeping he talked to Associations about were Pest and Diseases and their treatment this became part of his talk to Western Galloway in 1969 and Pests and Diseases and their Treatment was the title of the talk he gave to East of Scotland Beekeepers' Association, in February 1970.

Latterly, he was supported by his wife Margaret who controlled the projector and slides. Although on occasions the role was reversed and James controlled the projector and slides along with the occasional comments to enhance the talk.

Margaret was an accomplished beekeeper in her own right, was a keen photographer which enhanced her talks - one of which was "From Bee Boles to Smith Hives".

Some of the ladies examine the Observation Hive:—L. to R.—Miss Jane Stewart, Mrs Margaret Braithwaite, Mrs Beatrice Brown, Mrs Isobel Ross and Mrs Elizabeth Adams.

Fig 19.5. Mgt. Braithwaite with a group of Lady Beekeepers

Both he and Margaret attended the Apimondia Conference in Athens in 1979 and she wrote about it and their visit (A Post Congress Tour) to Crete after the meeting in the

Scottish Beekeeper in 1980. She also did Association presentations on the Apimondia Conference e.g. East of Scotland, in November 1979.

James was a great supporter of the Scottish Black Bee which allowed him to work a single Brood Chamber system and he over wintered on ekes he made and nadired under the brood chamber. Captain Thake followed his lead and did the same successfully.

James visited Craibstone in June 1984 collecting 30 grafted queen cells in a 5 framed queenless nucleus to produce and support Bernhard Mobus in the production and distribution of the Maud Scottish Black bees.

James successfully showed his honey. In 1968 he won the Sir William High Cup at the East of Scotland Beekeepers' Association annual Honey Show.

He was an SBA Expert beekeeper - he was awarded it in 1960, his wife Margaret and his brother George also were awarded the SBA Expert Beekeeper Certificate the same year.

He was Secretary of the East of Scotland Beekeepers' Association for over 16 years, President for 3 years.

He was a representative at SBA meetings.

He was a well known beekeeper both at home and abroad and was awarded the SBA Dr John Anderson Memorial Award in 1979.

The 1979 Scottish Beekeepers' Association AGM Report states "that President George Smith presented James Braithwaite, the Dr John Anderson Award, a certificate and a Caithness Vase for his many years of service to Beekeeping and the SBA."

When he died in 1990 Robert Couston said that " the Name of James Braithwaite will always be associated with the late David Robb... they amongst others pioneered the technique of producing cut comb ekes especially from heather sources... He was always willing to visit other associations to pass on his experience."

James and his wife Margaret attended numerous Apimondia Conferences making friends with beekeepers from all over the world and who he and his wife Margaret entertained when they visited Scotland.

George Halden Braithwaite was born in 1917 ten years after his brother James and died in 1999 at the age of 82. George, like James, as a boy helped his father keep his bees. However, unlike James who stayed in Dundee working in the family business and continuing to keep bees, George left the Dundee area, going south to London and the South of England, working with the International Commission distributing supplies from the USA during WW2. While carrying out his duties in Gloucester he found a hive of bees which he started to look after. This rekindling his interest in bees and beekeeping. On his return to Scotland (Forfar) after the war he decided he would start up a Commercial Beekeeping Business. In his first year he made up 150 hives -Smith Brood Chambers which he had bought as flat packs and floors. However, to cut costs he made the roofs and crown boards from Tea Boxes he acquired from the business in Dundee. He had apiaries around Errol and took bees to the heather at Glen Clova and had a mating apiary at Glen Queich. He had such a great success in his first year of honey production that he and June his girlfriend, were able to get married in 1950. Unfortunately, in 1952 he damaged his back while shovelling snow and so he gave up his dream of being a commercial beekeeper and joined James in the family business. Regarding Queen rearing George told an interesting story about buying an Italian Queen from a lady from Broughty Ferry. She had bought it and then did not require it. He duly bred from this queen only to find that bees from the second crosses had become extremely aggressive. After this he and James killed every yellow queen they found. This is probably the reason James was such a great supporter of the Scottish Black Bee.

During the years that followed he became an SBA accredited beemaster, expert beekeeper and finally honey judge. George would go on to judge at the Royal Highland Show in 1989.

Braithwaite and George Smith N.D.B. discussing this year's entries after the judging at the Royal Highland Show".

Fig 19.6. Photo of George Braithwaite as Judge at RHS

George was awarded the SBA Dr John Anderson Memorial Award in 1997. George served on the Executive of the SBA, was a life member of East of Scotland. Beekeepers' Association. George was a great supporter of the Dundee Flower and Food Festival and helped organise the East of Scotland Beekeepers' Associations stand and the honey show at the Food and Flower Festival.

George wrote about Ekes and Cogs for Cut Comb as well as how to make Association Show schedules clearer for exhibitors and judges. He did talks to Associations as well as doing apiary demonstrations. George like James was an enthusiastic beekeeper who was always encouraging and helpful to his fellow beekeepers.

Fig 19.7 George Braithwaite at Honey Show stand

On his death his wife June bequeathed £5,000 to the SBA to be used for Educational Purposes.

When the Scottish National Honey Show in conjunction with Ayr Flower Show was cancelled due to financial reasons it was June Braithwaite who got Dundee Council to hold the Scottish National Honey Show in conjunction with the Dundee Flower and Food Festival (Show). Unfortunately, due to the impact of Covid the Dundee Food and Flower Show no longer exists and the Scottish National Honey Show is now run in conjunction with the SBA Convention.

George and James were not the only beekeepers to use Braithwaite Tea Boxes. Neil Anderson who produced and circulated the plans for the hive designed by Willie Smith and which Anderson called the Smith hive - started his beekeeping with BTB hives the BTB standing for Braithwaite Tea Box Hives. His girlfriend (and later his wife) at the time also kept bees in BTB hives. It was only after his visit and work experience

(around 1942/3)with Willie Smith,

(arranged by his father who knew Willie Smith's neighbour in Innerleithen) that in 1945 he changed to Smith hives. (Neil Anderson had wanted to go into the beekeeping business in a big way and wanted to meet and get advice from WW Smith)

The timber of the Braithwaite Tea Boxes was made to make up National Hives . I am not sure to what extent BTB hives were adopted within the Dundee area and how easy it would have been to acquire the tea boxes to make the hives. Since it was during WW2 it may have been due to a shortage of timber and the greater availability of Braithwaite Tea Boxes that the tea boxes were used.

Recently the Braithwaite family gave East of Scotland Beekeepers' Association the vase that James Braithwaite was given by the Scottish Beekeepers' Association in 1979.

Fig 19.8 Caithness Vase presented to James Braithwaite - photo by W Hunter

East of Scotland have still to decide what they will do with the vase and how James Braithwaite will be remembered and recognised through this vase.

Whatever they decide I am sure it will be a fitting tribute to James A Brathwaite and a lasting memory of a selfless, helpful and knowledgeable Expert Beekeeper.

Robert Couston 1922 - 1993

Fig. 20.1 Photo of Young Robert Couston

Robert Couston was born in Perth and went to Perth Academy. He served in the army in the Middle East during WW2. Was Assistant curator at Perth Art Gallery and Museum. In 1949 he left his post as Assistant Curator at the museum and became advisor in beekeeping with the East of Scotland College of Agriculture, retiring in 1987. Bob Couston died suddenly at home on 10th May 1993 just before he was supposed to represent the interests of Scottish Beekeepers at a meeting with The Ministry of Agriculture, Fisheries and Food organised in London.

He was awarded the Dr John Anderson Memorial Award in recognition to his contribution to Beekeeping in 1987. He was President of the SBA from 1982 to 1985, and was chairman of the Bee-farmers' Association 1979 to 80.

He had a National Diploma in Beekeeping.

He was a Fellow of the Royal Entomological Society, was a member of the Magic Circle and also enjoyed fishing.

Bob Couston wrote in the *Scottish Beekeeper* on a regular basis.

On the Golden Anniversary (50 years) of Dunfermline Beekeepers' Association, he wrote a section for their Anniversary booklet.

He wrote about his family links with Dunfermline and how his grandfather (whom he was named after) was given a desk from Andrew Carnegie - the richest man in the world at that time and who had lived in Dunfermline before emigrating to the USA. The article is worth reading not just on why Andrew Carnegie gave his grandfather the desk but also as it gives an insight into Bob Couston's role as a Bee Advisor.

He wrote two Books on Bees.

Fig 20.2. Cover of *The Principles of Practical* Beekeeping

The Principles of Practical Beekeeping - first published in 1972 with a facsimile copy published in 2012 which is still available to buy.

The book covers most aspects of beekeeping from a glossary of beekeeping terms, the life cycle of the honey bees, introducing queens, late summer management (including preparing bees for the heather) as well as a section on the enemies of bees.

Although this book was written in the early 1970s, before the threat caused by Asian Hornets and Varroa, this is a book that has a lot of information on bees and beekeeping and is just as useful today as it was when it was written. It is well illustrated with photographs showing how the particular manipulations are carried out (The photographs of the 2012 facsimile are not as clear as the 1972 edition but are clear enough to follow).

This is a book that is well worth having and will help every beekeeper in the care and management of their bees, dealing with swarms, rearing queens and producing and preparing honey for sale.

Fig. 20.3 Cover of *Fly with the Beeman*

His second book *Fly with the Beeman* was published in 1989. This book was his follow up book to Principles of Practical Beekeeping. Bob Couston felt that by writing a book of both personal and amusing anecdotal stories on beekeeping, it would help illustrate some of the practical aspects he had written about in his first book and thus give more experienced beekeepers additional information. The book is a number of short stories covering many aspects of practical beekeeping and it reflects his character and personality as well as him being a bit of a bon vivant, he enjoyed the odd drink and I have heard that he was known for coming home late and sleeping in late and being a bit of a rake.

When you read this book, you will smile a lot and on the odd occasion laugh out loud.

Bob Couston left all his colonies of bees to Willie Robson - he had planned to give him his bees even telling him to "Come and Collect Them" just prior to his death. Willie speaks about them as being good bees

Fig. 20.4 Photo of older Robert Couston

His sudden death in 1993 was a great shock and loss to Scottish Beekeeping.

The Glasgow Beekeepers

Ian Craig (1937-2018), Charles Irwin and Eric McArthur (1935-2021) - made a major impact in Scottish Beekeeping from the 1970s through to the late 2010s. much of their work and contributions can be read about in the Glasgow Beekeepers Association Centenary Book: *The Glasgow Beekeepers*. It covers most aspects on beekeeping and is a must have if you work a double brood box system or you are setting up an observation hive.

All three were individually awarded MBEs for their contribution to beekeeping.

Unlike Willie Smith whose bees went to George Hood or Robert Couston whose bees went to Willie Robson and were an addition to the colonies they already had, Ian Craig's bees on or just before his death in 2018 or should I say the majority of his colonies went to a beekeeper in the North East of Scotland who I am told instantly re-queened with Buckfast queens. Ian was proud of his bees and that they produced on average over 100lbs of honey per year. It seems such a shame that a lot of his queens were destroyed when so many people could have benefitted by getting a queen which would have helped to improve their stock.

Luckily not all his bees went to the North East and his bees still live on in the West of Scotland and the Kilbarchan Area through the Association at Kilbarchan and his family.

Fig.21.1 Photo of Ian Craig

Ian was a teacher and he passed on his knowledge of bees and beekeeping through the SBA education system (Ian was SBA Education Convener for a number of years) both in training and through the exam system, through talks at Association meetings and on Apiary visits.

If you are interested in working a double brood management system you should read the section in the *The Glasgow Beekeepers*. There are also sections on reading the hive and queen rearing.

He was President of the SBA from 1997- 2000 and 2006 - 2009. In 2001 he was awarded the Dr John Anderson Memorial Award and awarded his MBE in 2013.

Eric McArthur   1935 - 2021

Fig. 21.2 Photo of Eric McArthur

Eric was a writer -he wrote prolifically both in books and magazines, he was uncompromising in his views and always put bees first. He was a great supporter for the Scottish Black bees and he felt strongly about pesticide and the treatment of Varroa making AGMs and the odd Association meeting/ talk very interesting.

He was editor of The Scottish Beekeeper for 10 years.

He was awarded the SBA Harry Brown Award in 1983 and the Dr John Anderson Memorial Award in 2004.

Eric died shortly after being awarded an MBE in 2021.

Charles Irwin

Fig. 21.3 Photo of Charles Irwin

Charlie is a well read and very knowledgeable beekeeper. His handling/manipulation of bees is the best I have seen and his phone number is in the contact list of most beekeepers in the Lanarkshire area. His impact through the observation hive at

Kelvingrove Art Gallery and Museum is immeasurable.

For many years with Peter Stromberg and Ian Craig, he held / taught sections of the Glasgow Beekeeping Beginners Course for those interested in beekeeping in the Glasgow and surrounding areas. Charlie has been a great ambassador for Beekeeping in the Lanarkshire and Glasgow area as well as for the SBA and Scotland.

He was awarded the SBA Special Award in 2015 for his services to the SBA and as the SBA Insurance Convener. He was awarded an MBE for Services to Beekeeping in 2019.

If you want to learn more about Ian, Eric and Charlie and the contribution they have made to Scottish Beekeeping read *The Glasgow Beekeepers*. The book covers a lot of what they did and covers almost all aspects of beekeeping. It is a wealth of beekeeping information.

Andrew Mitchell Abrahams

Fig 21.4 Photo of Andrew Abrahams

Andrew Abrahams was born in July 1948 in the West of Scotland (Dunoon). He came from a farming background. He graduated with a BSc (hons) in Agriculture and then taught abroad for a few years before becoming involved in the seafood industry which led him to taking up Oyster Farming in 1976 on the Island of Colonsay (An island in the Inner Hebrides on the West Coast of Scotland which is only accessible by boat). When he moved to Colonsay he was neither interested or had been previously involved in keeping honey bees. In the late 1970s Andrew was looking for other income streams to supplement his earnings. He had found / become aware of a number of old hives and nucleus boxes in a barn in neighbouring farm buildings. These hives had been used by Joseph Tinsley who had been the Head of Beekeeping at the West of Scotland Agricultural College, and wrote the Books: *Beekeeping Up-To-Date* and *Practical Advice to Beginners in Beekeeping*. From 1946 to 1966 Colonsay and Oronsay (The adjacent island which connects with Colonsay at Low tide) was used as a queen rearing and breeding station by Tinsley for *Apis mellifera mellifera* (The indigenous Black bee). The island was isolated and gave Tinsley control over the breeding of the bees and allowed him to improve and protect the sub species. Tinsley was worried about the introduction/ importation of other strains, races/sub species of honey bees into Scotland, particularly the Italian sub species and the hybridisation and the impact that this was having on the native Scottish Black bees. Tinsley had seen the sub species of Italian Bees being improved and becoming more popular, he therefore felt there was a need to improve the native black bee and to protect it.

He hoped that through improvements the benefits of keeping Black bees would be greater and the race /sub species would become more attractive to beekeepers with more keeping them.

On seeing the stored hives in the out-building Andrew Abrahams saw a possible business and financial opportunity in keeping bees. Bees had not been kept on Colonsay since Niall (Nially) McNeill left the island in the late 1960s. (McNeill had been trained by Tinsley in queen rearing and he had been involved in sending queens and nucleus colonies to Tinsley. When McNeill left he took the bees to Islay, leaving the hives not being used in the barn /out-building on Colonsay. (Andrew would initially use these double walled hives but moved onto single walled hives, for ease of working, later.)

To learn the basics and to gain experience in beekeeping Andrew worked for 2 summers with the bee farmer Athol Kirkwood at Heather Hills Honey Farm, Blairgowrie and helped to look after 1500 hives on the East of Scotland. During this period of time Andrew saw hundreds if not thousands of colonies, this gave him an insight into those colonies, those that were hybridised and those that were not. He would be able to use this knowledge when choosing his stock/colonies for Colonsay.

Another great influence on Andrew was Bernhard Mobus. Bernhard suggested to Andrew that he select native /locally adapted colonies of honey bees that would be able to survive the harsh conditions /weather of Colonsay - bees that had adapted to the Scottish harsh climate. Bees that could forage at low temperature and fly in strong winds as well as be able to get mated at lower temperatures. Mobus believed that the Native Black bees were best suited for Colonsay, having adapted to such poor weather conditions in Scotland over centuries as well as having previously survived on Colonsay i.e. during the Tinsley's queen rearing and breeding project. Andrew Abrahams therefore re-introduced Black bees to Colonsay, selecting 14 colonies - 2 colonies from Strathhardle, 4 from the Fintry Hills and 8 from the Ochil Hills and 16 queens (ten queens from Bernhard Mobus and his Maud Strain of Black bees and 6 queens from the Galtee Black Bee Breeders in Ireland).

Andrew selected colonies on observation, ones that did not show any signs of hybridisation.

Andrew evolved a system of bee management that was linked to the available bee forage as well as working to maintain the gene pool and thus the sustainability of this isolated population of Apis mellifera mellifera .

Andrew felt restricted to keeping less than 100 colonies due to the limited pollen available on Colonsay.

Andrew believes that pollen is very important to bee health and that poor pollen availability and pollen from only single floral/crop sources are detrimental to bee health making the honey bees more susceptible to disease. His aim was to ensure the bees had sufficient pollen along with a variety of pollen available throughout the year to maintain good bee colony health. Andrew therefore restricted the number of honey production hives to around 64 along with around 32 wooden and polystyrene 5/6 frame nucleus hives to support sustainability and self sufficiency. Andrew was also aware that he needed to have a genetic pool sufficiently high enough to stop in breeding and the problems that this creates. He wanted to avoid bringing in additional queens and bees. He wanted to maintain sufficient colonies to maintain the gene pool and not reduce/diminish this pool over time. Andrew believed he needed to keep more than 60 but less than 100 colonies with a genetic variation from the initial 30 queens (30 queens x 10 mated drones = 300 variations) to protect the gene pool. So, he maintains 8 apiaries with approximately 8 hives and 4 nucs. at each apiary, to maintain the balance in protecting the genetics and in maintaining good bee health through good nutrition from the limited pollen and nectar sources available on Colonsay. This meant he would not have to bring in bees that could introduce pests and diseases as well as the possibility of bringing in bees that were not pure Apis mellifera mellifera. (Colonsay bees are free from the Foul broods and Nosema although chalk brood is present. Also, because the bees have been on an isolated island since the late 1970s the bees are Varroa free and therefore do not have the issues Varroa brings).

The colonies / hives are not moved through the season and the position /site of the apiaries are chosen because of the shelter they offer and the forage available throughout the year. Andrew marks and clips his queens in April, and works his honey producing colonies – Queens which are one or two years old are kept in double brood chambers adding 2 or 3 supers above a queen excluder through the season. This helps to manage the swarm impulse during the honey producing season. Colonies are checked for signs of swarming by looking for queen cells from early to mid-June until the end of July (The bees on Colonsay swarm much later due to the difference in climate compared to other parts of Britain). To check, the double brood chambers are split and the bottom of the frames of the top chamber and the top of the frames of the bottom brood chamber are looked at for queen cells, reducing the disruption to the bees in the process. Andrew harvests the honey in September leaving approximately 30lbs per hive for over wintering.

3 or 4 year-old queens are kept in single brood chambers and are used to produce splits or are Demaree'd to produce queen cells and or splits. These 3 or 4-year old queens often producing supersedure queen cells which are used to make increase.

His aim is to conserve and improve the population of black bees he keeps.

He breeds for good productivity of honey, gentleness and lack of swarming.

He uses 5/6 frame nuclei and apideas to ensure self-sufficiency of his isolated bee population.

He has approximately 4 nuclei per apiary and thus ensures sustainability in each apiary.

When he makes up the nucs. he does 4 frame splits sometimes using frames from different hives to ensure he does not weaken any of his stocks/ colonies. The new nucs. are then moved to another apiary on the islands to stop drifting of bees either back to the original hive or to a nuc. that has the original queen.

Benefits of using nucleus hives are:

1. You can carry out a Demaree for swarm control and stock increase.
2. To draw out foundation - get a better frame
3. As a brood factory - to give to other colonies frames of brood to boost them
4. Use the nuc. to take frames of brood from strong colonies to slow down the strong colonies' development.
5. To have a source of queens throughout the year to replace for example drone layers.
6. Excess nucs. can be sold and are therefore a source of income.
7. Nucs. can be used for honey production where 5 or 6 nucs can be moved easily to take advantage of a honey flow where hives are not sited.

Andrew uses Apideas to get his queen mated quickly and to take advantage of the limited opportunities / windows of good mating weather/conditions he has on Colonsay. He sometimes overwinters queens in 10 or 15 framed apideas, he can do this due to the mild weather on Colonsay due to the Gulf Stream.

He rears small numbers of queens / queen cells regularly throughout the active season so that it is not an issue if the queens cannot get mated due to poor weather.

He sells over 1 ton of honey a year through sales both locally on Colonsay and by mail order.

He sells queens and nucleus colonies of bees.

He runs beekeeping classes.

He provides bees / material for research (Including *Braula coeca*) as well as disseminating queens and bees and therefore their genetics through Scotland and even further afield.

Andrew has made his beekeeping business a great success.

DNA analysis have shown that the *Apis mellifera mellifera* genetics of his isolated population is very good i.e. pure *Apis mellifera mellifera*.

More recently the importance of conserving the gene pool of the pure races /sub species of honey bees and the need to protect this genetic material for future survival of the species has been identified.

In 2004 Andrew attended the SICAMM Conference in Laeso, a Danish Island off the West Coast of Sweden and which also has a population of *Apis mellifera mellifera*. In 1992 Denmark passed a law that Laeso was a protected area for Black bees. Other beekeepers on the island were not happy and took the case, to overturn this law, to the European Court, but lost. However, these beekeepers continued to illegally keep hybrid bees on the island, even importing bees and in the process introduced Varroa and Acarapis mites onto the island.

The Conference in 2004 was to focus attention on the need to protect the Black bee population on the island – one of the few places (Like Colonsay) in Europe to have pure Apis mellifera melifera. At the conference Andrew met Flemming Vaesgnes who told Andrew that he needed to ensure his bees on Colonsay were protected for the future and that legislation was needed, to stop something similar to what was happening on Laeso, to be put in place. Flemming reiterated his views to Andrew at Apimondia in Ireland in 2005.

Andrew spent a lot of time looking at ways to introduce legislation to protect the Colonsay bees.

Unfortunately, there were issues round both Agriculture and Conservation legislation which made it impossible to give the bees protection. However, because of the devolved Scottish Government and in particular the Wildlife and Natural Environment (Scotland) Bill 2011, Andrew with the help of Peter Peacock a Labour MSP along with the help and support of other influential people and organisations, were able to create an opportunity to persuade and introduce legislation that would protect the bees.

In 2013 Andrew Abrahams achieved reserve status for the colonies of *Apis mellifera mellifera* on Colonsay and Oransay under The Beekeeping(Colonsay and Oransay)

Order 2013 as part of the Scottish Wildlife and Natural Environment Act 2011.

In recognition for his work / contribution to beekeeping Andrew Abrahams was awarded the Dr John Anderson Memorial Award in 2024 by the Scottish Beekeepers' Association.

Andrew continues to work on a plan to ensure the future and the sustainability, maintenance and protection of the Honey Bees on Colonsay and Oronsay. I have heard on the grapevine that plans are being put in place in conjunction with the Scottish Beekeepers' Association to make this happen and that when Andrew retires that his work will continue and that the bees will be looked after and their future assured. The importance of conserving this genetic pool and the contribution made by Andrew Abrahams in making this happen cannot be over emphasised. It is also important we see what happened on Laeso and make sure nothing like that happens on Colonsay. The bees must be protected for the future - we have a responsibility to support Andrew and make sure this happens we cannot let the Colonsay Bee Reserve be compromised.

And what of the other beekeepers of today, who will stand the test of time. From past history it is the recipients of the SBA Dr John Anderson's Memorial Award who will be the most likely to be remembered.

And what about the significant beekeepers of the past I have not written about, like the Bee historians Rev. J. Beveridge.

Fig. 22.1. Photo of Rev J Beveridge

Rev J Beveridge, who had a great library of bee books which he gave to St Andrews University, Dr Malcolm Fraser who wrote A. *History of Beekeeping in Britain* and whose wife bequested the J M Fraser Memorial Fund or Dr Tennent

Fig 22.2. Photo of Dr Tennent

(Who left a bequest - the Dr Tennent & Mrs Tennent Memorial Fund, where the income from the fund is for the benefit of the Moir Library) or Thomas Baillie who contributed to the development of the SBA during WW2 or the Honey farmers like A C Kirkwood and Willie Robson or even the great exhibitors of Shows like Mr Foubister of Alford.

Fig. 22.3 Photo of Mr Foubister

Past SBA editors like Robert Skilling or authors such as Margaret Anne Adams who recently wrote the book *Pollen Grains and Honeydew - a Guide for identifying the plant sources in honey*. It will be for someone else to write their stories and to give them a voice that will echo down the years.

A Brief History of Scottish Beekeeping and Beekeepers

www.ingramcontent.com/pod-product-compliance
Lightning Source LLC
Chambersburg PA
CBHW041243240426
43670CB00024B/2967